PHOENIX SCIENCE SERIES

Chapters 1–4 originally appeared in *Social Behavior and Organization among Vertebrates,* edited by William Etkin and published in 1964 by THE UNIVERSITY OF CHICAGO PRESS

Library of Congress Catalog Card Number: 67–28654

THE UNIVERSITY OF CHICAGO PRESS, CHICAGO & LONDON
The University of Toronto Press, Toronto 5, Canada
© 1964, 1967 by The University of Chicago. *All rights reserved*
Phoenix Edition published 1967
Printed in the United States of America

SOCIAL BEHAVIOR

FROM FISH

TO MAN

SOCIAL BEHAVIOR FROM FISH TO MAN

by

WILLIAM ETKIN

with a chapter by

DANIEL G. FREEDMAN

Phoenix Books

THE UNIVERSITY OF CHICAGO PRESS / Chicago and London

Preface

The study of the social life of vertebrates has been one of the most exciting and productive fields of zoological research in the last quarter century. The revolution it has produced in our understanding of the animal mind has aroused a great deal of interest among students of the human behavioral sciences. Clinical and social psychologists, psychiatrists, and anthropologists have discerned in this new knowledge suggestive insights into the basis of human sociality. In the present work, four chapters from a larger cooperative text, *Social Behavior and Organization among Vertebrates,* edited by the present senior author, have been assembled and combined with a chapter by the second author, a clinical psychologist turned evolutionist. The first four chapters develop the principles of animal social behavior from the ecological and evolutionary viewpoints. The final chapter attempts to suggest some implications of these principles for the understanding of the behavioral system of man. The material is presented from a non-technical viewpoint and should serve to introduce the non-zoologist to this field.

It would perhaps be well to indicate at this point the distinctive orientation of the zoological student of behavior as contrasted to that of academic American psychology. The zoologist studying behavior is primarily concerned with the role it plays in the adaptation of the animal to its environment. Such naturalistically oriented study of behavior is now called ethology (from *ethos:* habits or customs). Ethology asks how the behavioral repertoire helps in the survival of the individual and the species. It asks these questions not only of the adult organism but of all phases of the life cycle and under all environmental variables, including season, population density, and so on. Ethologists bring their problems into the laboratory for critical analysis, but without losing sight of the adaptive value of the behavior studied. For example, it is not considered feasible to understand rat behavior in the laboratory unless one is fully aware of the rat's natural behavior in the wild. Thus learning, long the dominant concern of experimental psychology in America, is viewed by ethologists as *part* of the mechanism of adaptation, and its variations are seen in relation to the species specific problems each animal faces in its own particular mode of life.

Ethologists do not underestimate the enormous difference between human and animal behavior patterns, even though elements of cultural transmission are acknowledged to exist in animal societies. Rather than accept these differences simply as given however, the ethologist attempts to see human behavioral systems from the comparative, evolutionary, and adaptational points of view. As such, ethology brings fresh viewpoints to an understanding of the phenomenon of man. The final chapter of the present work will serve to point up the significance for human social behavior of the ideas developed in the first four chapters. We hope it will inspire students of human behavioral sciences to keep abreast of the continuing developments of ethology and to use them to broaden the base of their own thinking.

W. Etkin
D. G. Freedman

Contents

WILLIAM ETKIN

1

Co-operation and Competition
in Social Behavior

Natural Selection

DARWINIAN CONCEPTS

The triumph of the theory of evolution and its Darwinian explanation
in the late nineteenth century had repercussions in almost every field
of thought. Not the least of these followed the application of the
Darwinian ideas of competition and struggle for existence to the social
life of man. Economists, sociologists, and social philosophers sought
to explain social forms and practices in terms of their contributions
to the success and survival of the group. In academic circles, such
thinking led to the theories of the Social Darwinists. One of the main
conclusions of these theorists was that it is necessary to permit the
strong to exploit the weak, for by this action human progress is produced. As we look back now, we may perhaps view these theories as
attempts to justify the harsh attitudes of laissez faire capitalism toward
the social evils of the day (Hofstadter, 1944).

Whereas we will not be directly concerned with the development
of sociological theory in this book, we must take notice of these ideas,
because both their direct effect and particularly the reaction against
them strongly colored contemporary thought on social behavior in
animals. On the one hand, the significance, and even the relevance,
of animal studies to an understanding of man has been denied in the

effort to reject the fancied consequences for man. On the other hand, great efforts have been made to counteract the nineteenth-century interpretation of Darwinism at the animal level by denigrating the role of natural selection. We shall see that both viewpoints suffer from an inadequate appreciation of the complexity of animal social life. For a proper appreciation of the evolutionary aspects of our subject, however, it would be well to begin with a brief consideration of some of the biological fundamentals involved in our present understanding of natural selection.

The concept of natural selection, popularly called survival of the fittest, has had a stormy history since Darwin's day but has emerged clarified and greatly strengthened in contemporary biology. It may be briefly characterized as follows. The reproductive capacity of any species is so high that the increase in population tends to outrun the available necessities such as food and shelter. In the resulting competition some members of the species, being better endowed by hereditary characteristics, survive and reproduce more than the others. Because of their success the hereditary factors or genes which they carry tend to be preserved and passed on in greater and greater measure to succeeding generations. Through this process, the pool of genes characteristic of the species tends to shift in the direction of greater adaptability. A species long established in a particular environment reaches an optimal balance among its genetic factors, but with each alteration in the environment the balance of these factors in the population tends to shift. It is important to note that the competition upon which natural selection acts is chiefly that between members of the same species, because members of different species, having different modes of making a living and occupying different ecological niches, do not compete as directly with one another.

In the latter part of the nineteenth century many biologists doubted that natural selection was adequate to explain the facts of evolution. In fact Darwin himself was not exclusively committed to natural selection as the driving power of evolution. In particular he often fell back upon the inheritance of acquired characteristics for an explanation of specific instances of evolutionary change. This concept, sometimes referred to as Lamarckianism, after Lamarck, the eighteenth-century French zoologist who espoused it, maintains that modifications in the body tissues which are acquired during an animal's lifetime induce tendencies for similar modifications to be transmitted to its offspring. In relation to behavior, the idea of inheritance of acquired characteristics suggests that habits learned by experience in one generation tend to become "built into" the germ plasm as "instincts" in subsequent evolution.

Contemporary biology, however, has discarded the Lamarckian concept. For one thing, experimental tests of its claims have been overwhelmingly negative. For another, the development of modern genetics has given us clear insights into the mechanisms of inheritance, and, as these are now understood, they do not provide any way in which environmentally induced modifications in the body could influence the genes. Finally, biologists today find no need to fall back upon the Lamarckian hypothesis to explain evolution. According to our present understanding of the genetics of animal populations, the gene pool of a population is a balanced system wonderfully sensitive to selection pressures. The double set of genes in each individual (diploid condition) allows the accumulation of gene mutations in the population. These furnish the raw material upon which natural selection acts. Sexual reproduction permits rapid diffusion of genetic change throughout a population. Together these factors make natural selection speedier and more efficient than anything previously suspected by biologists, even early in the twentieth century. As a consequence, the old objection that natural selection is too slow and ineffective to account for evolutionary change has lost its cogency. Experimental tests have supported the modern view that natural selection operating on the genetic system of higher organisms can effect rapid and delicate adaptation of these organisms to the changing demands of the environment (Dobzhansky, 1951).

The above viewpoint is not controverted by the finding that other factors, such as group isolation and random variations in small populations, are also of some importance in determining evolutionary change and may, on occasion, lead to developments not fundamentally adaptive in nature. These factors operate as modifying influences to the driving force of natural selection and, whatever their importance in minor evolutionary change, do not displace the long-term adaptive nature of the evolutionary process as produced by the action of natural selection. In conclusion we may therefore say that the contemporary view of evolution is that it is guided and directed by natural selection. In the large view the characteristics of organisms must be expected to be adaptive in the sense of contributing to the long-run reproductive efficacy of the species as it lives in its own particular ecological niche.

SOCIALITY AND DARWINISM

Before we examine the implications of natural selection for social behavior, it would be well to clarify some terms. By "social response" we shall mean a behavioral response made by a group-living animal to another in its group but not to animals or objects which are not members of the group. We shall regard groups as "social" when the members stay together as a result of their social responses to one another rather than by responses to other factors in their environment. Groups that are held together by responses to such other factors will be called "aggregations." Thus a flock of sheep is a social group, since it is maintained by the social responses of the animals to one another; but the massing of insects around a light at night is an aggregation, since it results from their common attraction to the light. Of course, in many cases we are unable to decide, because of lack of appropriate evidence, whether a group is truly social or not, and the noncommittal term "group" may then be used.

Granted that modern biology accepts the primary importance of natural selection in evolution, the question remains: "What is it that has survival value?" It is obvious that what may be called aggressive potential, that is, the ability to win in a conflict with competitors whether by wit, bluff, or brute force, is often the deciding factor in survival. Yet if survival were possible only for the most aggressive and self-seeking individuals, it is difficult to see how social life could develop far among animals. But despite this difficulty social life is evidently very common among higher animals, particularly verte-brates. This is impressed upon us if we merely think for a moment of the large number of words that our language provides to describe the naturally occurring groups of higher animals. Thus of birds we say a flock of robins, a gaggle of geese, a covey of quail, a bevy of pheasants, a brood of ducklings, a hatch of chicks, a set of swans. Of mammals we say a flock of sheep, a herd of elk, a pride of lions, a litter of puppies, a pack of wolves, a farrow of pigs, a colony of monkeys or gophers, a school of whales.

In a book on social life in animals in 1927, Alverdes summarized the knowledge at that time of the occurrence of group life among animals. He classified and described at some length the kinds of social organization found among animals. From this work we gather an impressive view of the varied and widespread character of animal sociality. Indeed, if we take a literal view of what constitutes social behavior as defined above (i.e., distinctive responses between members of a group which tend to keep them together) and therefore include in it the interaction of the sexes in reproduction, we may regard social behavior as universal among vertebrates, since all reproduce sexually. But aside from the special question of sexual behavior, it is still evident that there is much more social life to be found in the animal world than the theory of natural selection, with its emphasis on competition, might suggest.

It is clear that the well-established social group affords many advantages to its members. When a good food source is found by one member, all are attracted to it by the behavior of the finder. Animals in groups are much less subject to predation than are animals in isolation. The individuals warn one another of approaching danger so that it is very difficult for a predator to get close to a group (Fig. 1.1). Group action against a predator, as we find it in the "mobbing" of a hawk by song birds and in defensive formations in ungulates, provides protection for the members. The musk oxen form a defensive

FIG. 1.1. When the pronghorn antelope detects danger, the hairs of the white rump patch are erected, producing a conspicuous white patch said to be visible for miles on the open plains. When one such animal dashes away, the flashing white signal alerts the others even if the herd is widely scattered. Such protective devices greatly increase the security of the herd. (After E. T. Seton, *Lives of Game Animals*, Vol. 3; by permission of Charles T. Branford Co., Publishers.)

ring with large males on the outside. Group formation, as in close flocking in birds is so effective in protecting the individual that, generally speaking, predators try to isolate an animal from a group before attacking it (Fig. 1.2). Group life also facilitates reproduction in many ways. Some of these, such as the bringing together of the sexes at the appropriate time, are obvious, but some which we will discuss later are rather subtle.

Undoubtedly, therefore, genes which make for co-operative behavior convey some survival value to their possessors by encouraging group living. Many naturalists and philosophers in the post-Darwinian era emphasized this fact. They tried to show that not only the competitive survive, but that co-operative behavior, too, is found among

BEFORE   AFTER

FIG. 1.2. A flock of starlings before a falcon appeared (*left*) and after (*right*) is shown here. The tendency of group-living animals to close ranks when danger appears is widespread throughout the animal kingdom. When the prey is in close order the falcon cannot attack them, since it would be in danger of injuring itself. (After N. Tinbergen, *Study of Instinct*, 1949.)

animals and has survival value. The Russian naturalist Kropotkin wrote the most notable of the books along these lines, *Mutual Aid* (1914). In it he collected instances, often anecdotal, of acts of co-operation and of individual self-sacrifice among animals and argued strongly that co-operative behavior is an important factor in evolution.

In spite of the earnest and well-meant efforts to show that co-operative behavior has survival value, the main force of natural selection must be expected to favor self-seeking, "anti-social" actions by the individual. This clearly follows from the point made above that natural selection operates chiefly in the competition between members of the same species and particularly between individuals closely associated with one another. In such competition the survivors are inevitably those individuals which gratify their needs at the expense of other members of the group or species. It was difficult for theoretic evolutionists to see how co-operative behavior could develop very far in the face of this situation.

A factor introducing some confusion into our thinking in this area arises in the special case of social insects which are well known to show self-sacrificing behavior on the part of individuals. The worker bee that stings an intruder dies as a result of her action taken in defense of the hive, since her entrails are pulled out with the sting. However, in insect groups such as ants and bees where reproductive activity is confined to one pair in the colony, the queen and her mate, the action of natural selection is very different from what it is in vertebrates. ⌈In these insects the co-operative behavior of the workers not only facilitates group survival but also insures transmission of the genes that promote this behavior. Such behavior preserves the queen, who carries genes making for co-operative behavior in her germ cells and who passes these genes on, not only to her workers, but to her queen daughters as well. Thus the bee that sacrifices itself in stinging an enemy of the hive actually helps insure the transmission of the genes which favor such behavior.⌋

But this very example indicates the weakness of the position of those who argue from the behavior of social insects to that of social vertebrates. For, in vertebrates, ⌈reproduction is a function of all members of the group.⌋ In contrast to the situation in insects, an individual which sacrifices itself in protection of the group insures that, in general, its own genes will *not* be passed on to the next generation. On the whole then, one would expect from such reasoning that the evolutionary pressure for "selfish" behavior would be considerably stronger than that for co-operation and that co-operative behavior would be much less conspicuous than its opposite in vertebrates.

But whichever of these views one might favor on theoretic grounds, it is clear that this type of speculative analysis leads to no definite conclusion. It is difficult to reconcile the action of natural selection with the widespread occurrence of social organization in vertebrates. The explanation for group formation offered by Alverdes was that social animals have a social instinct absent in non-social creatures. We shall see that this idea of a social instinct hides a multiplicity of interacting factors, many of them dependent upon experience rather than being innate as implied by the term instinct. It will be our task in this book to analyze the many factors which modern study of animal behavior has uncovered, especially in the last quarter of a century or so. We shall see that older ideas of the action of natural selection were hopelessly naïve and hid much of the subtlety and variety of animal life under vague generalization. For example, later in this chapter we shall see how aggressiveness itself has been so modified by social hierarchy and territorialism as to reinforce rather than disrupt social life. In the light of modern studies, the social life of animals is seen to be vastly more complex and intricately balanced than envisaged in the discussions of post-Darwinian naturalists. Before we examine hierarchy and territory, however, we must consider an aspect of co-operative behavior brought out in the experimental work of recent years.

Physiological Effects of Aggregations

In the second quarter of the present century, an experimental search for the basis of animal sociality was vigorously pursued by W. C. Allee and his students (Allee, 1951). They investigated the physiological effects of grouping among various non-social animals. This approach was suggested by the fact that such animals are often found in temporary aggregations in nature. These workers uncovered a number of favorable "group effects." Some of these involve a chemical or physical change in the environment produced by the members of the group. For example, if aquatic animals, such as goldfish or flatworms, are grown in unfavorable natural waters or water into which small quantities of harmful chemicals have been introduced, they do better when combined into small groups than as isolated individuals. In particular instances the improvement has been found to result from the neutralization of poisons by the body secretions of the animals. In other cases the loss of salts to the water by animals living in it was the basis of the beneficial effects observed. Water which has been improved by organisms living in it is said to be "conditioned," and its use in fish culture by aquarists has become a common practice. In other instances, as, for example, in mice raised in a cold environment, the advantage of group living has been traced to the heat and shelter provided by the animals' own bodies. In still other examples the beneficial group effect has been found to be based on behavioral or psychological characteristics, as in the case of the reduction in metabolism in snakes that are allowed to aggregate as compared to the same animals kept apart (Fig. 1.3). Members of a group were also found to stimulate one another to greater activity because of a tend-

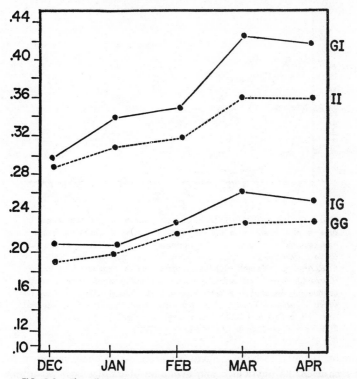

FIG. 1.3. The effects of aggregation and isolation on the rate of oxygen consumption of brown deKay snakes is shown here. One set of animals was kept in a group and tested in the grouped condition (GG) and also tested in isolation (GI). Another set was kept in isolation and tested grouped (IG) and again in isolation (II). The results show that animals tested while grouped (IG and GG) show lower oxygen consumption than the isolated animals, irrespective of the previous condition of the animals. Ordinates represent cubic centimeters oxygen per gram weight per hour. Abscissae represent monthly averages. (After J. Clausen, Cell. and Comp. Physiol., 8 [1936].)

ency to imitate (social facilitation). Experiments demonstrate such social facilitation in respect to eating, resting, and learning activity in animals as diverse as sunfish (Fig. 1.4) and children (Allee, 1951).

Perhaps the most biologically significant finding in these studies is the evidence that isolated cells give off substances into the medium in which they are grown which act as stimulants to the growth of other cells. This phenomenon has not as yet been adequately clarified, and the reality and nature of the hypothetical substances is still in

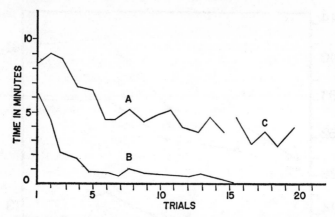

FIG. 1.4. The social facilitation of performance in a simple maze by green sunfish is shown in the graph above. Curve A shows the average time required to pass through the apparatus by thirty isolated normal fish. Curve B indicates the time required for thirty animals grouped three at a time. After the fifteenth trial the grouped animals were tested as isolates (Curve C). Since these animals showed the same performances as those continuously isolated, it is clear that their previous grouping had not resulted in greater learning. The facilitation must therefore have resulted from the leadership furnished by the fastest animals in the groups. (After E. B. Hale, *Physiol. Zool.* 29 [1956].)

doubt. If found to be of general occurrence, this phenomenon might be basic in the physiology and evolution of multicellular organisms.

From this work we can see that there are sometimes subtle selective benefits that accrue to animals by virtue of being aggregated into groups. We can appreciate, therefore, that there are evolutionary pressures, previously unsuspected, for the formation of social groups among non-social animals. Yet though such factors may have played a part in the origin of social life among animals, it is doubtful that they are strong enough or sufficiently general to be of great significance in explaining the existence of the more complex social organizations. In any case, though this work shows us something of the complexity of the factors involved in the survival value of group life, it tells us nothing of the physiological mechanisms which operate in maintaining the integrity of social groups.

As an example of how little we understand of the physiological aspect of group life in the less highly developed societies, we may briefly consider schooling behavior in fish. This is a very widespread phenomenon among fish, particularly such as live in the open sea—

herring, for example. Schools of fish generally consist of animals of similar size, often all of one species. Yet we know little about the benefits this schooling habit affords the animals. Many suggestions have been made: that the animals are protected from predators by the confusing effect of their great numbers, that they economize in energy required for swimming by keeping in the slip-stream of the nearby fish, etc. But these have not been experimentally demonstrated. We do know from experiment, however, that some schooling fish are visually attracted to any moving object that is neither too large nor moving too rapidly but are repelled by the same objects at close quarters. These responses probably account for the regular spacing of the fish

FIG. 1.5. Jacks schooling in large marine tank. Note the uniformity of spacing maintained by the fish. (Drawing adapted from a photograph, courtesy of Marine studios.)

in the school (Fig. 1.5). But the pattern and mechanisms of these responses have not yet been extensively analyzed because of the obvious difficulty of experimenting on animals that normally live in the open sea (Shaw, 1962). The role of social facilitation and other factors in the flocking of birds was emphasized by Crook (1961).

Individuality and Dominance

INDIVIDUAL RECOGNITION

Schjelderup-Ebbe, in his studies of bird flocks (see his summary, 1935), made the fundamental observation that the members of small flocks of many species of birds recognize one another as individuals. This observation was the outgrowth of the fact that his study of their group life was intensive enough so that he himself was able to recognize the individual members. This basic methodological insight—that the experimenter must, by means such as artificial marking and others, identify the individual in a group in order to understand its behavior —has become the basis of a host of studies that generally confirmed and extended Schjelderup-Ebbe's conclusion, which we may quote in his own colorful language:

Every bird is a personality. However, the word "personality" must not be understood to have the same meaning in this statement as when used in regard to human beings. What is meant here is that any one bird, irrespective of the species to which it belongs, is absolutely distinct in character and in the manifestations of character from any other bird of its species. This may sound odd, but that is only because the individual and social psychology of birds has been regarded too superficially. No attempt has been made to know each individual bird in a given flock. So to know them, however, is the most important prerequisite for the full understanding of the general and comparative psychology of birds. The ability to distinguish each individual provides the key for the solution of a series of problems which we should otherwise be unable to solve and which are not only of ornithological interest but also of importance for the understanding of the general continuity which prevails in all life.

This characteristic of individuality in animals, upon which, as seen above, Schjelderup-Ebbe placed the greatest emphasis, is strikingly manifested in many aspects of group behavior. For example, in a huge colony of gannets, gulls, or other sea birds, each individual recognizes its own mate and behaves quite differently to it as it approaches the nest than it does to other members of the flock. To the human observer the birds may look and sound alike, yet such a bird shows recognition of its mate when it is fifty or more feet away. In parent-offspring relations, individual recognition is often strikingly shown (although in the commonly studied laboratory animals, the rat and mouse, it is markedly absent). In most herd animals the mother-child relation is highly specific; in the fur seal, for example, among the welter of young moving about on the beach, a mother returning from a feeding expedition of several days unerringly picks her own child and will give suck to no other (Bartholomew, 1953).

Schjelderup-Ebbe's studies, however, were largely confined to the competitive aspects of individual recognition in birds. In this connection, individual recognition gives rise to the phenomenon of behavioral dominance or, to use Schjelderup-Ebbe's term, "despotism." It is perhaps unfortunate that the prominence given to dominance behavior in recent studies has tended to obscure the more general significance of Schjelderup-Ebbe's contribution in respect to individuality. Though we shall have repeated occasion to refer to the individuality of members of an animal group in other connections, we are here concerned with it, as Schjelderup-Ebbe was, in connection with intra-group competition. Studies along these lines have been carried out most extensively with birds, particularly with domestic chickens. Our initial comments will be based, therefore, largely on these studies.

HIERARCHY

In a small group of hens (up to about ten members) which have been living together with ample space for some time, very little aggressive behavior is ordinarily seen. Chickens are commonly fed by scattering the grain widely among them or using long feeding troughs, and each animal obtains a reasonable supply. If a competitive situation is set up by placing a single small pile of grain before such a group when hungry, an entirely different picture appears. One of the hens, which we will call the alpha or No. 1 animal, immediately approaches and feeds actively. Some other individuals may stay near the grain but make little or no effort to share in the food. If some of the grain becomes scattered, animals nearby may eagerly peck at it. Should the alpha animal notice them, it may make a threatening gesture or may peck the interloper. The latter immediately shows fear, retires, and never returns the peck. If the alpha animal is removed from the scene by the experimenter, another hen immediately takes her place and apparently by common consent is allowed to dominate the scene in the same way. Repeated removals of the top animal show that all the animals were eager to feed and were well aware of the food but were prevented from doing so by the individual at the food. This animal, called the dominant, was able to take precedence at the food and by its presence there prevented the other members of the group from even attempting to feed. Thus the group is seen to be composed of an array of individuals which form a hierarchy of precedence. Such precedence in competition assumed by an individual and acquiesced to by other members of the group is called "behavioral dominance." In stable chicken groups (Fig. 1.6) the order of dominance is usually clear-cut and forms a linear hierarchy in which alpha dominates all

	Y	B	V	R	G	YY	BB	VV	RR	GG	YB	BR
Y												
B	22											
V	8	29										
R	18	11	6									
G	11	21	11	12								
YY	30	7	6	21	8							
BB	10	12	3	8	15	30						
VV	12	17	27	6	3	19	8					
RR	17	26	12	11	10	17	3	13				
GG	6	16	7	26	8	6	12	26	6			
YB	11	7	2	17	12	13	11	18	8	21		
BR	21	6	16	3	15	8	12	20	12	6	27	

FIG. 1.6. Peck order as established by bringing twelve hens together is shown in the chart above. Each bird is designated by letters indicating the color code used to identify it. The number of times a given animal pecked other members of the flock is given in the vertical columns. The number of pecks and from which animals pecks were received are read in the horizontal column. It can be seen that each animal pecked only those below it in this liner hierarchy. (After A. M. Guhl, *Scientific American*, 1960.)

others; beta, all but alpha, and so on down the line, the omega animal being subordinate to the rest (Collias, 1944; Allee, 1951).

This pattern of precedence is found to apply to almost all competitive situations, at least in daylight when chickens are normally active. Thus the dominant animal assumes control of food, water, roosting places, choice of mate (by roosters, not hens), and any other competitive feature. The important point to recognize is that the precedence involved is not one which is fought over but is a regular characteristic mutually agreed upon and recognized by all members of the group. Occasionally a subordinate may overstep its bounds, and for such a social error it receives a severe peck. The subordinate does not peck back but flees. If in its flight it should encounter an animal subordinate to it, it may pass on the punishment by pecking. In a group of strange adult birds artificially brought together, this pattern of pecking is the most obvious expression of dominance, hence the term "peck order" often used by Schjelderup-Ebbe and later authors. Even aside from the fact that it is descriptive only of aggression in birds, this term is unfortunate, since it emphasizes the fighting that takes place. Such fighting, however, is largely an artifact resulting from the usual experimental conditions. We prefer the terms "rank-order" and "social dominance" for this discussion.

The induction of fighting by crowding enables the experimenter to determine dominance quickly, but from the theoretic viewpoint, the fighting involved is misleading. Natural or long-standing artificial groups that are stabilized and uncrowded show little actual fighting. The great importance of dominance in social life is that it acts as an organizing principle which minimizes aggression by, in effect, securing to the dominants the fruits of victory without disrupting group life by conflict.

Dominance behavior in chickens is clear-cut, vigorous, and quite uniform in expression, the alpha animal practically always dominating the rest. Such invariable dominance might be described as complete or absolute dominance; Schjelderup-Ebbe's term was "despotism." Allee used the term "peck-right." In other animals, however, dominance may be of a partial or relative character, called "peck-dominance" by Allee. In doves and pigeons, for example, no complete dominance of one animal over another appears; rather a count of pecks delivered and received may show a consistent balance in favor of one animal. On a statistical basis, pigeons in a flock may be arranged in a hierarchy of partial dominance, alpha delivering more pecks to beta or any of the others than he receives in turn and so on down the line.

Since partial dominance implies a failure of individuals to recognize one another and to accept social status once established, it is questionable whether it should be regarded as a part of social dominance in the sense of the term as used by Schjelderup-Ebbe. In any case, whereas absolute dominance is an organizing principle in social groups leading to a diminution of conflict within the group, partial dominance has little effect in this regard. It may be primarily a measure of the level of aggression in animals which do not form a well-organized group because of lack of facility in individual recognition.

Dominance hierarchies are not always linear. Especially when first formed they may be irregular, A dominating B, and B dominating C, but C dominating A. Such a triangular hierarchy tends quickly to break down into linear form in chickens but may persist for a long time in animals with partial dominance. Some animals may form hierarchies at two levels. All male chickens dominate all female chickens so that we may consider that there are separate male and female hierarchies. Prairie dogs similarly show a two-level arrangement. In mice under some conditions and in the Indian antelope in small herds,

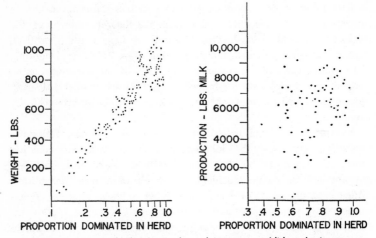

FIG. 1.7. Even such placid animals as dairy cows establish a dominance hierarchy in their groups. Dominance is established at about six months when the developing animals begin to take their pushing fights "seriously" after much play-fighting in younger animals. The "fights" are often mere threats by head movements or brief pushing so that these escape the notice of casual observers. Rarely are the animals so closely matched that real trials of strength occur. The plot on the left shows the high correlation between weight of animal and the proportion of the herd it dominates. Since weight correlates with age, the position of the cow in the hierarchy in these highly inbred animals seems determined largely by its birth order. As shown by the plot on the right, milk production does not correlate well with rank order. (After M. W. Schein and M. Fohrman, *Brit. J. Animal Behav.* 3 [1955].)

there is one dominant animal, all other members of the group being at one level in the hierarchy.

The vigor with which dominance is expressed may vary greatly in different species. Thus many herbivores may show only the mildest, if any, expressions of dominance in the grazing herd or may even feed from restricted bins without any evidence of precedence (Fig. 1.7). Yet in other species (the swordtail fish, ground birds like the chicken, carnivores such as the wolf, and some primates like the baboon and macaque monkey), dominance patterning of social relations is clearly important and strongly expressed, the dominant taking precedence in all competitive situations.

Besides species differences, another source of variation in the expression of dominance is that this behavior may be expressed in some aspects of life and not in others. Sexual behavior is most commonly the strongest area of the expression of dominance. Thus numerous ungulates (for instance, the red deer) show the strictest dominance among males in breeding season. The buck in rut collects and dominates a harem of does and rigorously excludes other males. Yet, out of rut during the major part of the year, the bucks live in loose herds showing little social interaction or expressions of dominance. Females, on the other hand, have well-organized, permanent herds showing clear-cut dominance patterning based on mother-offspring relationship with an old female as boss and leader of the herd. Almost no aggressive action need be shown in such a herd because each individual finds its place gradually as it grows up (Darling, 1937). Among Indian antelope studied in the zoo, there is one dominant male who asserts his dominance over other males conspicuously while on open range, but when the animals crowd around the feeding box, all evidence of his dominance disappears. He has to push his way into place as does any other animal. Thus, as in so many examples of animal studies, the behavior is "situation" specific. Dominance is commonly conspicuous in situations which are competitive in the normal life of the animal.

Dominance should not be confused with leadership. The lead in group movement often goes to the animals who are most alert to environmental change and, therefore, initiate the movement and are followed by the others. Where dominance is strongly and aggressively expressed, as in macaque monkeys, baboons, or the Indian antelope, the dominant males ordinarily do not maintain any sort of guard for the group and so do not initiate or lead. In the red deer, where the dominance of the older females is expressed more beneficiently, they maintain an alert and lead and direct herd movements. It is difficult to define the concept of leadership precisely enough for experimental analysis. In a simple situation Stewart and Scott (1947) found no correlation between dominance and leadership in a flock of goats.

An important aspect of dominance behavior in animal groups is the way it operates to make them closed societies. In a stable group, as natural groups generally are, each animal knows its status and keeps to it. If a stranger is introduced into such a group, the animal is evidently under considerable psychological disadvantage and is subject to persecution by all of the "natives." Thus in chickens, if a vigorous animal, say the alpha animal of one group, is introduced into another group, it first assumes a subordinate position, avoiding rather than retaliating attacks. However, as it becomes familiar with the situation, it may begin to fight back and gradually work its way up the scale to an appropriate position in the hierarchy. On the other hand, an animal not naturally so vigorous may be so beaten and oppressed as to succumb altogether with little show of resistance. Chicken farmers are familiar with this phenomenon and therefore will not ordinarily introduce a single individual into a strange flock. This rejection of the stranger (xenophobia) has been observed in numerous animals under field conditions. In the macaque monkey, for example, strangers, even females, attempting to enter a group are usually driven off by the males (Carpenter, 1942).

The establishment of dominance patterns under experimental conditions has often been studied to try to discover the factors involved (Collias, 1944; Guhl, 1953). When two strange hens are brought together, they immediately approach, watching each other closely. If there is a difference in size or vigor between them that is obvious enough to be detected by the human observer, the chickens can be counted upon to see it at a distance. In such a case one will, by lowering its head and avoiding the other, show that it accepts the subordinate position, whereas the other, holding its head high, shows its dominance by its freedom of movement. Again quoting Schjelderup-Ebbe (1935) for his colorful language describing his experience with numerous species of birds: "One bird, the subordinate, evinces apprehension, fear, and occasionally even terror of the other. The ruling despot discloses his identity by complete lack of fear toward the subordinate individual, and sometimes—strange to see—completely ignoring the other's existence."

If the birds being brought together are closely matched, they may approach, each one trying to keep its head higher than the other. If, by any accident such as stumbling, one head should be lowered, the issue is decided; that one becomes subordinate. In the usual experimental case, the animals are closely matched and approach close enough to fight. The fight is usually a matter of a few swift pecks, and one retires vanquished (Murchison, 1935). Such a decision is often seen under natural conditions where new relationships are being set up. In general, however, a fight occurs only when the opponents are so evenly matched that even to their expert eyes no other way of reaching a decision is apparent. In most cases the dominance-subordination relation is settled at the first encounter of the animals (Guhl, 1953). In many animals aggression takes the form of threat displays rarely involving real fighting (Fig. 1.8).

Once established, a dominance pattern tends to be highly stable. Some of the factors making for stability have been revealed by experimental and observational studies. For one, group-living animals have, in general, a rather good memory with respect to individual recognition within their own group. Thus a chicken removed for two weeks from a group can often be reintroduced without disturbance, since its status is remembered by all. After two weeks or a molt which

FIG. 1.8. This shows the fighting "dance" of two rattlesnakes. In their competitive aggressions these snakes attempt to push each other but do not strike. The fantastic figures assumed by such fighting snakes were once assumed to be courtship dances and are familiar to all in the caduceus symbol of the medical profession. (After H. Hediger, *Psychology of Animals in Zoos and Circuses*, 1955.)

changes their appearance, the recognition tends to be lost. The reputations of parrots and elephants for memory of individual associates, including human beings, extending over years seems to be well authenticated.

A second factor in stabilizing dominance organization is that the behavior of both dominants and subordinates tends to reinforce their status positions. Thus the dominant animal in most forms in which complete dominance prevails is readily recognized by the way it carries itself. As in the quotation from Schjelderup-Ebbe above, the bold look in the bird, the free movement, and indifference to the other animals characterize the dominant. Contrariwise, the manner in which the subordinate carries itself and avoids the dominant betrays its status. In some animals (for instance, the baboon and macaque) in which dominance is harshly expressed, each dominant male tends to occupy a semi-isolated area since other members of the group (except females in heat) tend to avoid it. The posturing "advertisement" of dominance in the Indian antelope is carried to an extreme. He assumes an awkward, stiff, strutting walk, holding his head high on stiffly bent neck, with his ears folded tightly flat against the neck. In many familiar species of deer and in many other large mammals, the rutting male shows similar awkward poses. Clearly then, many dominants maintain their position by a special conspicuous carriage, which seems to gain in conspicuousness by its non-adaptive character. Thus the folded ear of the antelope does not point toward the source of a sound and is in marked contrast to the mobility of the ears of the other animals in the group.

The importance of this factor of display is shown by the phenomenon of prestige as a factor maintaining social status. The dominant hen, for example, maintains its position of dominance as it gets old, weak, and less agile, even while losing its vision. It may then make even more of a show of its dominance display, carrying it so far as to seek occasion to chase or peck at subordinates. Eventually, however, one or more of the subordinates, despite its conditioning to subordination, begins to show signs of revolt. At first, perhaps, the revolt is shown only by making pecking and threat gestures in the direction of the dominant when the latter is turned away or is at a safe distance. Gradually the boldness of the subordinate increases until it eventuates in the inevitable battle and quick defeat of the former despot. An old and infirm despot when deposed in this way is indeed a sorry sight, becoming utterly dejected and full of fear. In a chicken yard such an animal often dies within a few days, presumably as a result of the traumatic experience as well as physical injury inflicted by its former subordinates. The maintenance of social status by prestige as well as the extreme effects that follow its loss shows how important a role dominant-subordinate relations play in the lives of at least some social animals.

It would appear evident that since a stable dominance pattern reduces the amount of fighting that takes place in an animal group, it must help to make the group life more efficient. Such experimentation as has been done tends to support that concept. For example, a comparison was made between two groups of hens allowed to stabilize their social structure and two in which individuals were being interchanged frequently so that the social structure never could settle down (Guhl, 1953). The stable groups pecked less, ate more, gained more weight, and laid more eggs than did the others. When the welfare of the individual is considered, it is also found that, in flocks kept on short rations, the dominants secured more food than did the subordinates; indeed, top dominants gained weight while subordinates starved. It would seem that not only may dominance lead to survival for the individual under conditions of stress, but it may also help in group survival. It is well known that among ungulates such as deer and bison a great increase in population may lead to overgrazing of the area in which they live. Then if a particularly hard winter comes along, the entire herd may be lost because of the lack of food, since such grazing animals lack any well-developed dominance organization with respect to food competition. However, in animals with dominance in feeding, the animals higher in the scale secure all available food, whereas the subordinates are starved out early. Since environmental stresses such as food supply and weather are frequently controlling factors in the survival of animals, it is evident that the advantage enjoyed by dominants enables them and, with them, the species to survive. It would appear, therefore, that there are strong evolutionary pressures making for the development of dominance as a principle of social organization. Whereas it is certainly not universal among group-living animals and in some species takes exceedingly mild and limited forms, it is a principle which nevertheless helps stabilize many types of social groups. At the same time, by eliminating many of the harmful effects of aggressive activity expressed within a group, it permits evolutionary pressures to push the development of the capacity for aggression to a high degree. Thus dominance patterning of social organization is clearest in those group-living animals which show a high level of individual aggression.

Dominance hierarchy is by no means the only factor in the control and modification of aggression in the social life of animals. In later chapters we will discuss the role of courtship, parental care, early experience, and learning in relation to aggression. At present, however, we wish to turn to territoriality because this phenomenon is very closely connected with dominance in control of aggression in animal societies.

Territoriality

In an attempt to see this phenomenon in its most general aspects, we shall define territoriality as any behavior on the part of an animal which tends to confine the movements of the animal to a particular locality. This definition differs from those conventionally used for reasons which will become apparent later in the discussion.

Territoriality is a phenomenon of overriding importance in the lives of some animals, particularly birds. Although occasionally noted by ornithologists earlier, it was not until 1920, when the English student of birds Eliot Howard published his book on territory in bird life, that the significance of the phenomenon was really appreciated. Perhaps the most famous American study is that Margaret Nice made of the song sparrows in the Cleveland area (Nice, 1937). From this study we take the following account which illustrates territorial behavior in a representative song bird.

About half of the males and a small proportion of the females remain in residence during the winter, the others migrating south. Each resident bird tends to confine his activities to a restricted region but, particularly in bad weather, may join with others to form a winter flock. In the early spring, often as early as January if there is a warm spell, each resident male begins to show special behavior in relation to its particular area, a plot usually of about one acre in extent. Here he takes up a conspicuous place on a reed or tree and sings loudly and persistently. This behavior becomes more and more marked as the season progresses until in March the animal stays permanently in its own territory and sings frequently from his "headquarters" or display station. Should another song sparrow alight anywhere near, the resident assumes a very watchful and aggressive attitude which is usually sufficient to drive off the intruder. In some cases the intruder, usually a migrant arriving from the south, stands his ground and shows that he intends to stay by remaining in spite of the threats of the territory holder. In this case a rather definite behavioral interchange or "territory establishment ceremony" ensues. The newcomer puffs out his feathers, often holds one wing fluttering straight up, and sings softly. He appears subordinate to the confident dominant resident. The latter, now silent, watches him closely for a while as he flies from bush to bush. Then he gives chase. The intruder flies off but persistently returns and eventually turns on the resident, and a fierce pecking fight on the ground ensues. The battle is brief; the intruder, if defeated, flies away. If, however, he holds his own, the birds separate. Each, retiring to a different high point, sings loudly. It is thereafter noted that the original territory of the resident is now divided into two territories. Each bird keeps strictly to his own territory, clearly recognizing and respecting the invisible boundaries separating it from that of its neighbors. Each bird frequently sings from his display area in his territory and behaves as a self-confident dominant, driving off any small bird transients and fighting any male song sparrow intruders. When the females arrive from the south, pairing takes place in the male's territory, and the male stops singing for a month or two. The nest is built there, and both birds largely confine their activities to this region. The male remains active in defending his territory; although after the spring migration is over, there is much less occasion for fighting and territorial defense behavior slackens.

The above example is typical of territoriality in its classic form

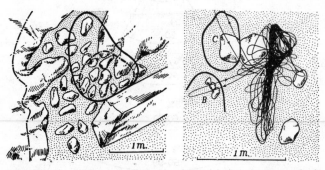

FIG. 1.9. The Japanese ayu is a stream-living fish that maintains a feeding territory usually around selected rocks in the stream bed (*left*), the central area of which it defends from other ayu. When crowded, those animals, unable to maintain a territory, live in non-territorial aggregations. At right is a tracing of three minutes of an ayu's life showing how its movements are confined to its territory except for two excursions made in driving off interlopers. (From Miyadi, *in* A. A. Buzzati-Traverso (ed.), *Perspectives in Marine Biology*, 1958.)

particularly as seen in song birds. It is a geographical area to which the animal confines itself and from which it excludes others, particularly members of the same species, except its mate; such a territory is a "defended area." Such defended territories are of widespread occurrence among vertebrates (Fig. 1.9), being especially common in bony fish, reptiles, birds, and some orders of mammals, particularly primates. Many students of animal behavior use the word "territory" to refer only to this type of defended area (Collias, 1951; Burt, 1943; Carpenter, 1958).

On the other hand, the locality sense of animals may take a variety of forms, some of which bear resemblance to this classic type but differ in that the animal shows no tendency to exclude others of the species from it. Thus many mammals, such as rats, mice, pronghorn antelope, and many deer, show a very strong predilection for remaining within a limited area with which they are evidently familiar (Fig. 1.10). Thus field mice in nature and rats in warehouses have been found to confine their activities to very limited space of a few hundred square yards or a few rooms. Hunters chasing pronghorn antelope in

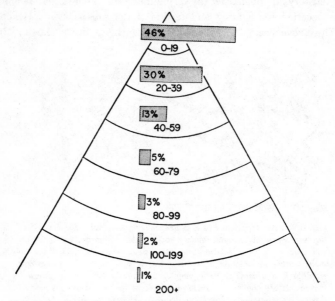

FIG. 1.10. The chart shows the distance in feet at which individual rats have been found (recaptured) after being released from the first point of capture in an urban site. This illustrates the strong tendency of rats to remain in the same area (home range); 89 per cent of the captures were made within 60 feet of the release point. (After D. Davis, *Quart. Rev. Biol.,* 28 [1953].)

our western plains often noted that it is impossible to drive an individual outside of the few square miles of area in which it is found. Even at considerable risk the animal persists in circling back to keep near to home. Such a sense of locality, which is characterized by the positive element of attraction to the area without the negative element of driving others off, may be called home range or home-range territory. Unfortunately, the distinction between defended and home-range territory cannot always be maintained in practice, since our knowledge of the natural behavior of animals under the varied natural conditions of their lives is often insufficient to enable us to say to what extent others are excluded from the territory. The unqualified term "territory" is therefore useful for such cases, and it can readily be qualified as home-range or defended as our knowledge justifies.

FORMS OF TERRITORIALITY

At best, the distinction between home-range and defended territory is not an absolute one, since it depends upon the very variable factor of aggressiveness shown toward others. Such aggressiveness fluctuates not only from species to species but within the same species from time to time and occasion to occasion. Thus, in the song sparrow discussed above, the males that remain in residence throughout the year tend to stay in the neighborhood of their breeding territory at all seasons. Yet in the late fall and early winter, they do not exclude other song sparrows from the area. In fact, the tendency toward exclusion of others is strongly expressed only in the early part of the breeding season, beginning usually in January or February. In June it wanes distinctly. During the molting period in September, the animal is very retiring and does not challenge trespassers. Later he may challenge adult males again to some extent but permit young males of the current year's brood to remain as long as they vocalize only with juvenile call notes and do not attempt adult song. In December the resident male neither sings nor challenges any others. Thus the male song sparrow maintains defended territory early in the breeding season and home range later.

Territorial behavior may be shown by a group, such as a herd or flock, rather than by an individual or pair. Such group territories are known to exist for many mammals and some birds. The red deer of Europe, for example, live in herds (Darling, 1937). Except in the rutting season, the males live in loosely formed herds separated from the better organized herds of females with their young. During rut, the adult males abandon their own territories and wander far and wide through the territories of female herds, seeking to round up as many females as possible and keep them together in a small group. The male rigorously excludes all other adult males from this harem area by constant patrolling and fighting when necessary. The hinds, however, do not share the stag's rut territory but tend to wander off unless persistently rounded up by the males. As a male is exhausted, he loses control of his hinds and returns to the uplands for rest. Territoriality in the red deer is thus complex, at one time being a group phenomenon, at another, an individual one confined to the male sex. Group territory seems to be of the home-range type. The rut territory is not a fixed geographical area but simply the neighborhood around the hinds as they move about grazing. Game cocks in a farm yard may similarly show a mobile territory, each cock staying with his hens and each such group avoiding the others so that there is no conflict. Yet each of these expressions shows a special behavioral relation the animals have to a special area.

Functionally, territoriality may be related to one or many of the fundamental life processes—reproduction, feeding, etc. As was obvious in the examples previously discussed, the typical defended territory seems to function primarily in relation to reproduction. Thus territoriality among song birds is most evident during the period of nesting. In an animal such as the song sparrow, since the mated pair remain within the territory during this time, it also serves for feeding, nesting, roosting, and raising the young. Such a territory may be characterized as a feeding-breeding territory. This, however, is not the case in other species. Many marine birds, for example, range over the ocean for feeding. Their nests are usually in great colonies along a favorable shore. In these colonies, territoriality is expressed very strongly with regard to a small area around the nest. Thus gulls, terns, gannets, and many other marine birds nest in large colonies in which each pair has a small nesting area, sometimes as little as a square yard or so, which is vigorously and constantly defended against trespassers, even young chicks from neighboring nests being excluded (Armstrong, 1947). Herring gulls pair up before adopting such breeding territories but conduct all their reproductive activities thereafter in the territory and desert it only when the young leave (Tinbergen, 1953). In other birds territoriality may function during only part of the reproductive process. Some male ducks, for example, maintain an out-of-the-way territory to which they conduct the female for copulation. A most interesting form of mating territory is lek formation, as shown by the European ruff, our prairie chickens, grouse, and others. In this type large numbers of males establish small individual mating territories in close proximity. The European ruff stands in his territory, watching to see that no other male oversteps its borders. When a female (reeve) appears, the males go into a frenzy of display, throwing the collar of long colorful feathers up around their neck to produce the conspicuous ruff and taking stiff "ecstatic" poses. The female chooses her mate without interference, and once she commits herself, the other males lose all interest in her movements (Armstrong, 1947). Members of a sage grouse lek not only display themselves visually but make their characteristic loud, booming sound. Since in these cases the female departs after mating and raises the young by herself, the territory is functionally restricted to mating. On the other hand, the winter territory of such birds as the mocking bird and English robin serves only in relation to self-maintainance and not reproduction (Lack, 1943). Group roosting territories are a familiar sight in winter for birds such as starlings. Bat caves may also be considered roosting territories, though, of course, mating also takes place there.

An interesting analogy to lek formation occurs in some fish, mostly stream-living. In the rainbow darter, for example, the breeding males congregate in shallow, gravelly places. Here the larger males stake out small breeding territories from which they drive all other males, maintaining a continuous guard. Females are permitted to enter, copulation and spawning taking place in the territory. In the red-bellied minnow, males congregate in a part of the stream appropriate for spawning. When a female swims in, males congregate around her, usually one on each side. Males compete for position around the female, but there is no fighting; and once two animals are lined up on either side, they are not displaced. Thus, in spite of the fact that many males are in the area, aggression is minimized, and two males co-operate from either side in fertilization.

RELATION TO AGGRESSIVE BEHAVIOR

A most significant aspect of territorial behavior from our present point of view is its relationship to aggressive behavior. The central point, of course, is that a territory of the defended type is maintained by the aggressive actions of the territory holder. Such action is directed essentially at "competitors," most often sexual competitors of the same species. Though the decision in a conflict over territory may involve actual fighting, even fatal combat, it is apparent that, once decided, the isolation of the aggressive individuals in separate territories eliminates further fighting. Thus territoriality minimizes the amount of actual harm done to members of the species by fighting. All types of aggressive behavior are not necessarily eliminated by territorial isolation. Many territory owners patrol the borders of their territories, singing persistently or otherwise displaying aggressively. Bulls of the fur seal and related species, for example, repeatedly charge to the edge of their breeding territories, making formalized open-mouthed thrusts at neighbors who reciprocate. Occasionally the charge takes the form of a belly slide that ends with the two animals snout-to-snout at the territory boundary, each "puffing at the other like a small locomotive" (Bartholomew, 1953) (Fig. 1.11). The absence of real fighting in the stabilized territory permits the reproductive life of a mated pair of song birds, for example, to be carried through without interruption by competing members of the species. In other animals, the fur seals

FIG. 1.11. The defense of territory by bull sea elephants involves loud trumpeting (the enlarged proboscis acts as a resonating chamber) and labored charging at rivals, each animal going only to the border of his territory. Thus neighboring males frequently end their charges in the upright posture as shown above. (From a photograph taken in the Antarctic by A. Saunders.)

for instance, life is not so quiet, since young bulls are more persistent in attempting to reach the females throughout the season.

Careful watch of the behavior of the territory defenders reveals that the relative tranquillity that often exists is dependent upon an important relationship between territoriality and dominance behavior. The well-established territory owner has an enormous advantage in any conflict with intruders. It can often be observed that a bird within its territory behaves and carries itself with the characteristic self-assurance of a dominant. But when for any reason it leaves its territory and trespasses on that of its neighbor, it assumes the lean look and furtive behavior of a subordinate. This is most dramatically seen in some fish which show their aggressive state by their pigmentation. Some male cichlids, for example, as they swim away from their territory, blanch noticeably in their nuptial colors. In many animals there seems to be a rough proportionality between the "self-confidence"— or tendency to dominate—and distance from the territory headquarters. Thus, when territorial manakins were experimentally subjected to competition for a stuffed female set down in their territories, each animal's courtship was expressed with a vigor that varied in proportion to the distance of the lure from the center of its territory. Similar results were seen with great tits, where the degree of dominance at an outside feeding station was found to vary with the distance from the animal's territory. This tendency for the animal to show dominance in relation to the site of the conflict in respect to its territory is seen in naturalistic observations of many animals. For example, in territory establishment in the song sparrow and in many territorial species of fish, reptiles, and mammals, intruders have difficulties in proportion to the smallness of the territorial possessions into which they attempt to insert themselves. This phenomenon generally places a limit on the reduction in size of territories; for when an owner's territory becomes small enough, he will fight to the death to prevent loss. This limitation of compressibility of defended territory generally has important ecological or practical consequences, as we shall see shortly. Here, however, we should note the resemblance to the behavior of a dominant in a group, who so often maintains a characteristic dominance isolation, as we mentioned above for baboons. An analogous phenomenon is the formation, within large herds, of subgroups around each male, as described for game cocks. From this point of view, territoriality may be viewed as only one, though perhaps the most definitive, of the ways in which a dominant tends to isolate itself or be isolated by the creation of a "no man's land" around it. Hediger has emphasized the importance of such "flight" distance relations between trainer and animal in circuses and zoos in maintaining the appropriate dominance relations between them (Hediger, 1955).

As much of territoriality is related to mating behavior, we often find that territorial defense is the function of the male only. This is particularly true of fish, like the sticklebacks, that build nests and of the lek birds. But in song birds, where male and female remain within the territory, the female may in the course of a few days gradually assume some defense activity, either against all intruders, as do the English robins, or against females only, as do some ducks and others. In some cases the roles are reversed; the females of the red-necked phalarope, for instance, hold the territory. The defense of lairs as territories by female mammals is common. The European rabbit defends the area of her nest irrespective of the presence of young. Within a male chameleon's territory several females may defend subterritories against one another (Greenberg and Noble, 1944). Winter territories may be held by female song birds such as the California mocking bird or the English robin. Thus, although defended territoriality is predominantly a male characteristic, it is by no means exclusively so.

THE ECOLOGICAL SIGNIFICANCE OF TERRITORY

As biologists, we are inclined to assume that any basic characteristic of organisms, behavioral or otherwise, is likely to have survival value. Hence, Howard, in his original discussion of territoriality, attempted to analyze the ecological value of territory holding. Many of the benefits seem obvious; yet since experimental work in this area is extremely difficult, it is impossible to be sure to what extent benefits that seem reasonable are really of importance. We must depend chiefly on an analysis of observations, especially comparative observations, of different species or the same species under differing conditions as a basis for our interpretations. Though some points of Howard's analysis have been strongly disputed—particularly the value of territory as an exclusive feeding range—for the most part his interpretation has stood up well and will be followed here.

We may consider first the relation of territory to reproduction. The male bird establishing his territory in the spring makes himself conspicuous. Song birds do this by loud and frequent singing, usually from conspicuous perches. In some species, the sky lark, for example, the male indulges in conspicuous flight over the area; in others, awkward poses are assumed, such as that of the European stork sitting on his nest klappering and waiting for a passing female to fly down to him. This self-advertisement presumably helps the female find unmated males. This is further indicated by the commonly observed fact that the males in many species become much quieter after mating. In lek birds, the mass effect of the advertisement in aggregation combined with the use of the same areas year after year likewise facilitates the finding of mates by females when they are ready. Though matchmaking may be important in some instances, however, it must clearly be a secondary consideration in typical song bird territory, since territory defense persists after mating, and some birds, for instance, herring gulls, pair before territory formation.

The persistence of territorial behavior beyond mating suggests that it is helpful in maintaining the association of the mates throughout the reproductive period. As we shall see in more detail in the discussion of reproductive behaviors, the co-ordination of the behavior of both mates during the long and complex sequence of activities that are necessary for effective bird reproduction is a major problem. By confining the reproductive individuals to a small territory, their behavioral and physiological co-ordination is facilitated. This would appear to be one of the most general, as well as important, functions of territoriality. Such facilitation may extend through to include parent-offspring co-operation. We shall discuss it further in Chapter 4 but here one aspect of co-ordination related to dominance ought to be mentioned. Dominants often interfere with mating among the subordinates in the group. Under experimental conditions, it is often found that subordinates among dogs and ducks and fowl often make no attempt to mate when in the presence of the dominant, although proving eager when the dominant is removed. In some species, such as the chicken and dog, where this suppression of mating is well marked and persists even in the absence of the dominant, the phenomenon has been called "psychological castration" (Guhl, 1953). In other instances (e.g., seals and rhesus monkeys), subordinate males in the group achieve mating only when the dominant is completely occupied by other females. Thus, by isolating each pair in an area in which they are effectively dominant, territoriality favors successful mating.

Another function which Howard saw in territoriality is that of maintaining an economic balance between food and population. In the song bird territory, he saw the establishment of freeholds wherein each family would find the food resources necessary for survival. Young birds have enormous appetites and high growth rates; nestling song sparrows increase their weight more than tenfold in the first ten days. At the same time, nestlings are delicate and must be brooded to be kept warm and protected from the weather. The presence of an uncontested food supply nearby is of obvious value. Yet the importance of this item, even in song bird territory, has been disputed. Food competitors of other species, for example, are not attacked by most species, and the intensity of territorial defense often drops markedly before the young are fledged, even though their food requirements are then at the highest point.

Even though the idea of territory as the freehold of an individual family does not seem to be as important as Howard thought, territoriality still seems to play a role in regulating animal economics in another way, namely, in acting as a population-regulating mechanism. Since territories are not indefinitely compressible, it is clear that, as population goes up, the range of the species is extended by the new individuals claiming territories peripheral to the group. In general, this entails less favorable areas. Song bird males that are unable to find adequate territories are generally unable to breed at all. Thus the population in a given area would tend to be stabilized around the number of families that can be supported by the resources of the area for that species. Some evidence that this is actually operative is seen in the fact that in some species the removal of one member of a territorial pair is followed, sometimes within a day or two, by the settling in its place of another bird (Lack, 1943). Such replacements presumably had been part of an excess of non-territorial, non-breeding birds. This phenomenon of ready replacement is by no means universal in song birds, and it may well be that in some species—such as the song sparrow, where it was not found—other factors may have been regulating population size. Perhaps, too, it varies with the year or the place. Nevertheless defended territoriality does seem to place an upper limit on increases in population. As a consequence, in species with this type of territory we do not find the tremendous variability in numbers from one year to another that is found in some other animals. In mice and rabbits, for example, which show home-range behavior, a tendency for population to run in cycles in some areas has been observed. The population builds up to a very high density in favorable years only to be cut down suddenly to a very low level. The lemming of Norway, a mouse-like animal, is the classic example of an animal whose population builds up to high density and produces a plague of animals. These move out of the mountains and keep migrating, eating farm produce, swimming rivers in their path, and pushing on until finally they reach the ocean into which they plunge to their death. Such cycles are well known among the Arctic rabbits of Canada, the lynx, and other predators that depend upon them. Such animal plagues of mice and rabbits, unlike the lemming plagues, often end with a great dying-off whose cause is obscure. Defended territoriality protects a species from such violent population fluctuations and illustrates the importance of the type of territorial behavior for an animal's general ecology.

Summary

We may summarize the conclusions we have arrived at in this chapter as follows:

1. Modern biology supports the concept that the primary driving force making for evolutionary change is natural selection. Therefore, all characteristics, behavioral as well as anatomical and physiological, must generally be adaptive—enabling the animal better to survive.

2. Social organization is often of great survival value for the group. Even at the most primitive level of loose aggregation, sociality may have decided advantages for survival.

3. However, evolutionary processes must be expected, at least in vertebrates, to favor competitive and aggressive behaviors on the part of the individuals. These would tend to disrupt social life. Since, in spite of this, group formation in vertebrates is very common, we are led to expect that there must be ways in which aggressive behavior is kept under sufficient control to prevent its interference with sociality.

4. Two factors which operate very widely in vertebrate social groups to control aggressive behavior are (a) dominance hierarchy based upon individual recognition among the animals of a group and (b) territoriality which confines aggressive and other behaviors to limited areas. These factors and their importance in the organization of social groups in vertebrates are described in some detail in this chapter.

2

Reproductive Behaviors

Introduction

The sexual mode of reproduction is almost, though not quite, universal throughout the organic world. Even among the simplest of organisms such as bacteria and viruses which had previously been supposed to lack sexuality, the essentials of this mode of reproduction are now known to exist. Modern genetic theory expounds the evolutionary advantage of sexual reproduction over the asexual mode. It thereby explains the ubiquity of this type of reproduction, in spite of the obvious behavioral complexity which it necessarily involves in requiring the co-operation of two parents.

It has long been obvious that sexual reproduction allows for the mixture of hereditary materials from two different sources. It thus permits new characteristics of selective value which arise anywhere in a species to spread throughout the population. By bringing together favorable characteristics which have arisen by mutation in different parts of a population, sexual reproduction increases the variability within a species and furnishes a broader base for the operation of natural selection.

The role of sexual reproduction in promoting diffusion of genes is still regarded as important. But another and even more significant relation of sexuality to evolutionary efficiency has been brought to light by recent developments in the field of population genetics. In higher animals and plants, sexually produced individuals have two sets of genes. Such individuals are said to be diploid. If the members of a pair of genes differ (heterozygous condition), one of them is generally dominant, the other, recessive. The dominant gene is the only one which is expressed in the characteristics of the heterozygous organism. Since recessives are thus "covered" by the dominant gene in the heterozygote, natural selection cannot operate to eliminate them in such individuals. Only the homozygous recessive shows the recessive characteristics, and only in these individuals is the recessive gene subject to the action of natural selection. As a consequence, recessive genes, no matter how deleterious, accumulate in the gene pool of the population. The level of each recessive in the gene pool is determined by the balance between the mutation rate which is giving rise to this gene and the rate of elimination of the gene by the action of natural selection on the homozygotes produced when two heterozygotes mate. Each species population, therefore, contains a store of hidden recessives in its gene pool. Should the environment of the species change in such a way that previously non-adaptive recessive genes now become advantageous, the species does not have to wait for the slow process of mutation to originate the now advantageous genes, for they are immediately available in the gene pool. Because of the survival rate of the homozygotes under the new conditions, the proportion of the gene increases rapidly. Thus the evolutionary process is enormously accelerated by the sexual reproductive mechanism. We thus understand why, in spite of the numerous difficulties it entails, sexual reproduction is almost universal among animals.

Patterns of Sexual Reproduction

LOWER VERTEBRATES

The primary behavioral problem of sexual reproduction is that of bringing the sperm and eggs together to effect fertilization. The simplest method—one widely used among aquatic animals—is the scatter method. The gametes (sex cells) are simply released into the water, the random movements of the actively swimming sperm bringing them into contact with the eggs. Although in some plants chemical attraction is known to guide the sperm to the egg, no such attractive substances are known to act over a distance in animals. Many behavioral adaptations have been developed, however, which increase the probabilities of successful fertilization. The common tendency for animals to form aggregations during the breeding period or to grow in dense clusters, as do shellfish and tunicates on rocks, is one of these. A more direct adaptation is the almost universal limitation of spawning to one specific and limited time period. The swarming of the palolo worms two or three particular nights in the year, related to the lunar cycle, is a classic example of such limitation. Such behavior may be regulated by response to environmental cycles of tide, temperature and light, etc., or involve inherent time sense in the animals. In the worm *Bonellia* and to some extent in the marine snail *Crepidula*, the larvae which settle down near females tend to develop as males; otherwise they become females. The male of a deep-sea angler fish attaches itself to the genital region of the female, developing there as a degenerate parasite with little more structure than necessary for release of sperm. In starfish, worms, and some species of marine molluscs it appears that spawning is co-ordinated by the stimulating effects of the sex products themselves. Thus once a few animals start spawning, a chain reaction is set up whereby all the animals in the area are induced to release their sex products at the same time. Males and females or their sex products are brought into propinquity in numerous other ways, thereby facilitating fertilization even though the germ cells are randomly scattered.

Far more common and important, particularly in the lives of higher aquatic organisms such as fish, are the social behaviors by which insemination of the eggs is insured. As a result of behavioral interactions called "mating behaviors," males and females come into close proximity. They then release their sex products into the water in a co-ordinate manner so that the spermatic fluid (milt) is spilled directly over the eggs. Mating behaviors are complex. They involve co-ordination of seasonal development so that both sexes are ripe and active at the same, appropriate season. The sexually mature animal must be capable of recognizing the species, sex, and exact status of sexual readiness of a prospective mate. The mating procedure itself consists of a complex series of movements bringing the sex openings of the mates close together and effectuating release of the sperm over the eggs as the latter emerge.

The intricacies of this behavior can be illustrated in the analysis of mating in the common leopard frog (Noble and Aronson, 1942). First the males and then the females of this species emerge from hibernation early in spring with their gonads fully ripe, the development of the gametes having been completed during the previous summer. The males have a strongly developed clasping reflex and attempt to grasp any appropriately sized object that moves before them. When clasped by another, a male frog or a spent female croaks vigorously, a female distended with ripe eggs remains silent. The croaking serves as a stimulus for the clasping male to release; otherwise, as with a ripe female, the male remains in the clasping posture (amplexus) indefinitely. When releasing her eggs, which she normally does only while in amplexus, the female executes certain shuffling and pumping actions. These appear to stimulate the male to match her activities by releasing a dose of semen directly over each group of eggs as they emerge. When completely spent, the female shows backward shuffling movements. The male then releases his hold, and the mating is brought to an end. With an early spring start, the larvae have adequate time to complete their tadpole stage and metamorphose by mid-summer.

The extrusion of the sex products into the surrounding medium, common in water-living forms, is of course impossible in land-living animals, since the gametes would be quickly killed by drying in air. Thus internal fertilization is necessary in land forms. The sperm are introduced directly into the appropriate ducts in the female's body in a process called copulation or coition. Many animals have an intromittent organ (the penis in mammals) by means of which the sperm transfer is effected. Some accomplish copulation without specialized intromittent organs. Birds, except for ducks and a few others, lack penile structures and accomplish coition by a quick apposition of the lips of their external vents or cloacae. Bird copulation has been described as a "cloacal kiss." The male salamander in many species, after a brief courting of the female, marches ahead of her and deposits the spermatophore, a gelatinous capsule containing sperm. The female picks up the spermatophore with the lips of her cloaca. Still more bizarre are the reproductive behaviors of some of the cephalopods (octopus) which release one of their arms specialized to carry a load of spermatophores. This isolated so-called heterocotylized arm is seized by the female and inserted into her genital tract.

The advantages of internal fertilization in terms of conservation of sperm and eggs and certainty of fertilization, are manifest. Furthermore, it permits two other advantages: internal development of the egg and the secretion of heavy protective membranes around the egg before its release from the female. It is therefore not surprising to find internal fertilization widespread even among fish. The sharks and the many viviparous bony fish practice it. Transfer is usually accomplished with the aid of modified fins called claspers or gonopodia.

Internal fertilization as such does not necessarily entail any greater behavioral complexity than does external fertilization with mating as discussed previously. The protection and care of eggs and young, however, which occurs in both animals with external and those with internal fertilization, adds an entirely new dimension of behavioral complexity to the reproductive process. The care of eggs often involves nest building. In some fish this may consist of little more than clearing an area in which the eggs may be placed. The jewel fish, for example, merely cleans off a spot on a rock before placing her adherent eggs there. The nest of birds such as the gannets and penguins may consist of little more than a cleared area or "scrape." On the other hand, not only may birds weave elaborate nests, but fishes may also construct such extensive shelters. The male of many species of nest-building fish constructs the nest. The stickleback, a small fish breeding in the weedy shores of streams and estuaries, does this with sticks and weeds cemented together by mucous cords secreted by his kidneys. In birds it is more often the female that does the actual nest building (Fig. 4.1), but sometimes the male contributes or, rarely, as in the phalaropes, carries on alone. Completely domed nests, even provided with runways leading to separate outhouse conveniences are

FIG. 4.1. The crested cassique, a bird of the American tropics, lives in colonies in which each nest is a long pendulous sack elaborately woven of plant fibers and lined with leaves and other debris. The nest is built by the female while the male watches. She first wraps long fibers around a suitable branch and builds downward, working largely from the inside. When the nest is finished, copulation and egg-laying take place. Then the male abandons his family. (From an exhibit at the American Museum of National History.)

built by some birds (Pycraft, 1914; Burton, 1954; Heinroth, 1958).

From the point of view of behavioral correlation, nest building raises many problems regarding the nature of the activity. How is the timing of the nest building determined? It must, of course, come after territory establishment and before egg laying. How are the correlations between the sexes in their contributions to these activities effected?

With or without nests, animals may guard their eggs. The stickleback male keeps watch over his territory, including the nest with its eggs, driving off all intruders. But most of his time is taken up with fanning the eggs. This he does by directing a stream of fresh water over them with his fins. The rate of fanning increases with the age of the developing embryos and presumably keeps pace with their oxygen consumption. Parental care such as this is by no means rare among bony fish, patricularly those that breed near shore and maintain individual territories (Aronson, 1957).

Among amphibia and reptiles parental care is quite rare, though it does occur. The male of the midwife toad of Europe carries the strings of eggs wrapped around his legs. For the most part, he hides in damp places at this time but will emerge to dip the eggs occasionally in water when dryness threatens; he finally seems to recognize the appropriate time to bring them to a pond to permit the developed tadpoles to hatch. Though reptiles have made the very important step in the evolution of reproductive mechanisms by producing the first truly land egg, that is, one able to develop without external source of moisture, they are not in advance of lower groups behaviorally. A few reptiles show some parental care: pythons coil around their eggs and keep them slightly warmer than they otherwise would be, and alligators are said to protect and supervise their brood after they hatch. Parental care, however, only rises above the level reached by bony fish in the birds and mammals. These two groups are of sufficient importance to require individual treatment.

BIRDS

Birds and mammals are warm-blooded (homoiothermal) creatures. Their bodies possess a thermostatic-regulatory mechanism controlling the production and loss of heat by the body. This mechanism maintains body temperature above that of the usual environment and keeps it constant despite fluctuations in the latter. The physiological advantage of homoiothermism lies in the greater metabolic activity possible at the higher temperature. The homoiothermal animal may continue its activity throughout the year, even in cold climates. But whatever its advantages, homoiothermism greatly complicates the problems of care of the young. The bird's egg, once it has passed an early stage of development, must be kept warm continuously by the parents; for, if chilled, the embryo perishes. This then requires incubation of the egg, a behavior which necessitates many adjustments. The parents must co-operate in some way to make it possible for at least one of them to incubate practically all the time. In pigeons, for instance, the female incubates from afternoon to the following morning, and the male takes over the rest of the time. Some passerine

FIG. 4.2. Post-nuptial continuation of courtship activities are common where both parents care for the young as in the gannets shown above. The gannets are seen with outstretched necks, clashing their beaks together like fencers crossing swords. This ceremony is repeated when one parent returns after being away on a feeding expedition. (After E. Armstrong, *The Way Birds Live*, 1943.)

and marine bird parents alternate in incubation for brief periods during the day, and one of them takes over at night. In some instances one parent (in the European robin, the female) does all the incubating. In any case, the problem of feeding the incubating parent remains. The male English robin brings food to his female, who leaves the nest long enough to partake of it. Where incubation is more evenly shared, food may still be supplied to the sitter by the mate, or each may forage for itself while off the nest. The parents must recognize and co-operate in relieving each other at appropriate times. Sometimes elaborate "nest-relief" ceremonies have developed in connection with the exchange of places at the nest (Fig. 4.2).

When the young birds first hatch, their feathering and physiological mechanisms for temperature control are not completely developed. Indeed, many birds (altricial) are hatched naked and quite unable to take care of themselves, and others (precocial), though feathered and able to feed and move about are delicate. It is essential for the physical protection of the young that incubation pass over into brooding. In brooding the parent tucks the young under her (or his) body and wings, thereby keeping them warm and also protecting them from rain or excess sun (Fig. 4.3). If naked young are exposed to the elements for too long (an hour may be too long in small birds such as wrens), they may die. Of course, effective brooding requires coordination between parents and young and between the two parents.

FIG. 4.3. The female in many birds as the red-backed shrike (*above*) protects the young from excess sun by shading them. The automatic character of this behavior in birds is illustrated by the fact that the female of one species has been seen to shade the nest even after the young were removed to the side. They then lay exposed to the sun in plain view of the female while she continued the shading behavior. (After E. Armstrong, *Bird Display and Behavior*, 1942.)

The brooding behavior is further complicated by the high food requirements and related factors. Young birds have an extraordinary rate of growth. The weight of young English robins was found to increase nine times in as many days. This high growth rate requires high food intake, in addition to the extra food necessary to maintain the homoiothermal condition, for energy. Many young song birds eat their own weight, or near it, in food per day. The feeding problem of a pair of song birds with a family of five is thus enormous. English robins have been seen to make, on the average, fourteen visits with food per hour, in one instance bringing over one thousand caterpillars to their young in one day (Lack, 1953). The young of other small birds (Fig. 4.4) were found to be fed three or four times per hour (Kendeigh, 1952).

Even after they are well grown and fully feathered, the fledglings are still a parental responsibility. Often the parents have to induce them to leave the nest, using "tricks," such as not feeding them, etc. But even out of the nest, the young are usually fed by the parents and only gradually become able to fend for themselves. During this period the parents, in so far as they are not preoccupied preparing another nest and brood, also guard the young, giving warning to them

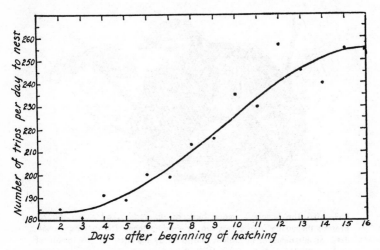

FIG. 4.4. The number of feedings the house wrens bring to their young increases as the young grow older and larger, requiring more and more food and less brooding. (After S. C. Kendeigh, *Ill. Biol. Monogr.* 22 [1952].)

by vocal signals of the approach of danger and in many cases attacking predators. Perhaps the most extraordinary parental protecting devices is the distraction display, or "broken-wing act," performed by many species of birds. When a predator approaches the nest or fledglings on the ground, one of the parents may rush in front of the enemy and flutter about on the ground with one wing awkwardly extended as though it were broken. As the predator is tempted to pursue this seemingly easy prey, the parent manages to keep just ahead of it and lead it away from the young, which meanwhile seek hiding places. Once while watching some newly fledged young robins hopping awkwardly on the ground, the author noted a cat approach-

FIG. 4.5. A common response of young birds to an alarm note sounded by a parent is to freeze in crouching attitude. Birds that show this response such as the stone curlews shown above are generally cryptically colored. As a result, such animals become almost invisible to the human, and presumably, to the predator's eyes. (Drawn from a photograph.)

ing stealthily. Rushing, as he thought, to the rescue of the young birds, he suddenly found an adult robin with broken wing fluttering frantically at his feet. Unthinkingly he reached to retrieve it, but it just eluded his grasp. Again and again for some twenty feet, he ran and reached for it, only to see the bird in the end dart off with perfect flight. Only then did he realize that he had been victimized by the broken-wing trick and led away from the young, which had meanwhile disappeared. We cannot now attempt to explain such behavior in objective terms but must be content to note it descriptively as an example of the extraordinary complexity of the behavioral adjustments between parents, both male and female, and their offspring (Emlen, 1955). (See Fig. 4.5.)

Compounding the intricacies of bird behavior is its co-ordination with seasonal conditions. In a general way, as we noted above for the frog, all animals breed only at appropriate seasons. But with the increase in complexity of reproductive life in birds, there is a corresponding increase in refinement of this co-ordination. Most birds of temperate climates migrate south when winter comes, and those which do not must be prepared for a period of great hardship in winter. An early start and a rapid growth rate during the first summer of life are of great advantage in preparing birds for their first winter. Yet, if the parents breed too early, there may not be enough insects or other food for the voracious young when they hatch. Consequently many birds must follow a precisely regulated schedule. Their complex pairing, nest building, copulation, egg laying, incubation, brooding, and feeding of young must not only be attuned to the regular cycles of the seasons but also subject to some further control to fit the vagaries of the weather in each season (Fig. 4.6). The northward-migration dates of many birds are quite regular to the calendar, some arriving on almost the same day year after year, and most of them showing but small variation in time. For example, in song sparrows near Columbus, Ohio: the first breeding males were found to arrive within a week of March 1, but the day varied within this period in conformity to the average mean temperature of the last ten days in February. Similarly, the first eggs were layed between April 15 and 23, and, within that period, laying was closely related to previous average

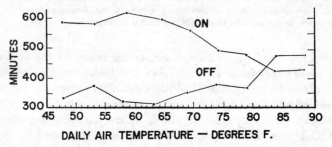

FIG. 4.6. The amount of time the house wren spends on and off the nest varies with the temperature, as shown above. Thus, incubation is much more persistent at low temperatures than at high. Above 80°F. the female spends more time off the nest than on it. (After S. C. Kendeigh, *Ill. Biol. Monogr.* 22 [1952].)

temperature (Nice, 1937). Even after being started, nest-building and other reproductive activities were interrupted or slowed down by a spell of bad weather. It is clear that in birds there is an overall mechanism that somehow regulates these behavioral activities in accordance with the season of the year; minor mechanisms further fit the behavior to the weather conditions of each particular year. All in all, the behavioral adjustments between female, male, young, and the environment in birds are so complex that they probably constitute the most highly elaborated reproductive behaviors found anywhere in the animal kingdom. Of particular interest for our later discussion is the fact that the successful raising of young birds depends upon long-continued co-operation between male and female. In song birds the sequence of territory establishment, pairing, nest building, mating, egg laying, incubation, brooding, and feeding of young is accomplished sometimes by one parent, sometimes by the other but in most species is shared by both.

MAMMALS

The basic biological organization of mammals is, in some respects, simpler and, in others, perhaps more complex than that of birds. We shall be concerned here only with the highest group of mammals, the placentals, since this includes almost all of the familiar mammals. The other groups, the egg-laying mammals and the marsupials, are but a surviving remnant of early types now largely confined to the Australian region. Placental mammals are characterized by the fact that the embryo develops within the mother's uterus, deriving nourishment and oxygen from the mother and giving up its wastes to her through a temporary organ called the placenta. At parturition (birth of young) the placenta separates from the uterus and is eliminated as the after-birth, which is generally eaten by the mother.

Though all placental mammals are born in post-embryonic stages, that is, with completely formed organs, there is considerable variability in the degree of development. Some, such as the bears and the primates including man, produce relatively small and helpless young that require many months of nursing and care before they can move about and find food for themselves. Others, such as the hoofed animals, porcupines, guinea pigs, and hares, are relatively far advanced in development at birth. Buffalo young can move along with their mothers within a few hours after birth, and hares can feed on solid food within a few days. Young porcupines become independent of their mothers in a week or so (Bourliere, 1954). In contrast to birds, many mammalian young develop quite slowly. In the larger land mammals nursing periods of a half to a full year are common. Bear young do not become independent of their mothers for two years. Mice require about three times as long as birds of comparable size to attain adult weight. In most ungulates sexual maturity is not reached for two or three years; in chimpanzees, for six to eight years; in humans, for about fifteen years; and in the elephant, not until about thirty years. During the nursing period the young are dependent upon only the female parent for food and general protection. Thus the initial dependence of the young centers around the female parent and, in contrast to birds, generally contains no role for the male. The long prepuberal developmental period affords extensive opportunity for learning and for the development of social relations between the mother and offspring.

The integration of the mammalian reproductive pattern into the seasonal and other environmental variables likewise is centered chiefly around the female. The larger mammals, particularly in temperate or arctic climates, have only one brood per year or sometimes one in two years (bears, for instance). The females and males are sexually active only during a limited period, the "rut" period. The males become active early in rut and maintain a more or less continuous sexual activity throughout it. In some mammals sexual activity is maintained throughout the year, and in many domesticated forms the limited rut period natural to the species tends to become continuous throughout the year. In such cases the males remain sexually active continuously during adult life.

In females the situation is more complex. There is cyclic activity, called the estrus[1] cycle, which consists of short-termed physiological cycles containing a brief interval of sexual excitement called heat or estrus. These cycles go on during the rut period but cease during the rest of the year. Thus sheep and most large ungulates show a fall rut period and are sexually quiescent at other times. In contrast to the males, which are continuously active during rut, the females accept the males only during their heat phase of their estrus cycle; if they are not impregnated at the first estrus, the cycle may be repeated at regular intervals during rut. In the non-rut period, estrus cycles cease. Such mammals are said to be seasonally polyestrus. Many tropical and domesticated mammals and some wild, temperate climate species run cycles all year long (permanently polyestrus). A few, such as bears and seals, have only one estrus period during the rut season (monestrus).

Some aspects of the estrus cycle must be mentioned briefly because they are important for an understanding of mammalian behavior. The term estrus (older spelling, oestrus) previously meant the period of sexual receptivity or heat in the female. Today, however, this is known to be correlated with changes in the entire reproductive system of the animal, and the term has been extended to include these changes as well as heat behavior. In a typical mammal, such as the guinea pig, the follicles containing the eggs ripen in the ovary during the early period of the cycle (proestrus), and the uterus undergoes certain

---

[1] The spelling "estrus" is used here for both adjective and noun.

changes. In estrus proper (the period of heat) or a few hours thereafter, ovulation takes place and, if coition has also taken place, the egg is fertilized in the upper part of the female ducts. In the next phase of the estrus cycle (metestrus), the embryo implants itself on the wall of the now fully prepared uterus. A placenta is established and pregnancy results. Should implantation of the egg fail to occur for any reason, the animal passes into a sexually quiescent stage (diestrus). After the diestrus rest period the animal runs another estrus cycle during the breeding season. If the non-breeding season meanwhile supervenes, estrus cycles cease altogether (anestrus). The rabbit and the cat and certain other mammals do not automatically run through the estrus cycle but remain in heat for some time until mated. During pregnancy the cycles generally cease except for traces. Many mammals also run one cycle after parturition, but during the period of lactation, cycles again cease. In some species even after an infertile copulation the female remains in a pregnancy-like state called pseudopregnancy, during which cycles are suppressed.

When the primate uterus which had been built up in preparation for an embryo breaks down after metestrus because of the failure of implantation, there is considerable loss of blood as the uterine lining is cast off. This is called menstruation. As can be readily understood from the origin of the bleeding, menstruation takes place not at estrus but rather between succesive heat periods in about the middle of the estrus cycle. In most primates the time of estrus can be identified by behavioral receptivity of the female and by structural changes in the sex system. It can thus be seen to alternate with menstruation. In the human, however, though menstruation is conspicuous, there are no definite estrus behaviors or morphological changes by which the time of ovulation can be observed, although it can be detected by refined electrical and temperature recordings. In summary, we may say that most mammals show estrus cycles characterized by definite heat behavior; primates, other than man, show both estrus and menstrual cycles; and man shows only menstrual cycles. Most of the fundamental physiological changes are similar in all three types. In a few lower mammals some bleeding from the genital tract may occur which is not menstruation. Dogs bleed slightly during estrus proper.

The nature of mammalian reproductive physiology as outlined above makes it clear that the burden of reproductive behavior necessarily falls primarily upon the female. It is her estrus behavior which determines the time of copulation, and she does not ordinarily permit

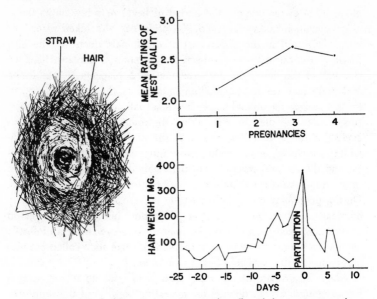

FIG. 4.7. Nest-building in rats is strongly affected by environmental temperature and by pregnancy. Both low temperature and closeness to the time of parturition increase nest-building. However, previous experience does not seem to be an important factor in these animals. Rabbits, on the contrary, improve the quality of their nest-building with the first three successive pregnancies, as shown in the graph, *upper right*. They line their nests with hair (*left*). This is pulled from the abdominal region where it becomes loose in the pregnant female. It is possible to quantify the changes associated with pregnancy by weighing the amount of hair which comes loose during a standardized combing procedure (graph, *lower right*). In the *left* picture the hair lining in the nest completely covers the young. (After Zarrow, Sawin, Ross and Denenberg, *in* E. L. Bliss, *Roots of Behavior*, 1962.)

copulation except in estrus. It is she alone who carries the burden of the developing embryo, and it is she with whom the young necessarily associate closely during lactation. If a nest or lair is prepared, the female builds it (Fig. 4.7). Only in a few species, as in the fox and the wolf, does the male contribute toward the care and feeding of the young. Generally, it is with the female that the young are associated, and from her they derive the basic mammalian social behaviors.

In contrast to the situation among birds, the contribution of the male mammal tends to be limited to that of insemination, and his social role is usually much reduced. His co-operation is generally not essential for the care of the young subsequent to fertilization. He commonly does not associate with the female and her young in any permanent manner. In some mammals, such as the cat and bear families generally, he is excluded by the female from contact with the young, being treated as their most dangerous enemy. We shall see that many of the behavioral characteristics of the mammalian male are related to the basic biological position of dispensibility after insemination. Of course there is much variability among mammals, but it is only exceptionally that the male finds an important niche in the social relations of parent and child among mammals. From the point of view of natural selection, mammalian males may be characterized as expendable. This characteristic helps to account for the kinds of specialization in behavior and structure that we find in many mammals (Bourliere, 1954).

Mating and Courtship Behaviors

GENERAL CONSIDERATIONS

The insemination of the female by the male, especially in land forms, often involves complex behaviors upon the part of both mates. Behaviors that directly lead to the transfer of sperm from male to female are called "mating behaviors." The commonly recognized behaviors of mounting, pelvic thrusts, and intromission of the penis in mammals or the "cloacal kiss" in birds are clearly to be included in mating behaviors. But in addition to these, we very commonly see many other activities between mates that do not seem to be directly involved in mating. For example, a male tern brings a fish to his "intended" mate. At this time the mate is perfectly capable of fishing for herself and this attention seems superfluous. After toying with the fish for a while the female may discard it. Clearly, hunger played no role here. Yet all along the shore in the spring, terns may be seen presenting pieces of fish to others and doing it, furthermore, with elaborate, precise, and formal ritualistic movements. Such activities between mates which do not directly help in insemination we shall call "courtship behavior." Of course, the distinction between mating and courtship behavior is arbitrary, and many behaviors are difficult to classify as either, since mating and courtship grade into each other. For example, during mating the male cat seizes the female by the scruff of the neck with his teeth. Is this behavior directly or indirectly involved in insemination? It would be pointless to try to distinguish mating from courtship here.

We will not, in fact, find it useful to try to maintain a distinction but will discuss both together in this section.

It is to be noted that the word "courtship" is used here in a much broader sense than is common in speaking of human behavior. For one thing, it is not confined to the male sex but may be conspicuous in female behavior. In the second place, animal courtship may continue after mating and be part of the characteristic activities of the two parents toward each other. Terns, for example, when they come back to the nest to relieve their mates in incubation, commonly bring a fish and present it with much bowing and posturing. Since this is clearly much the same behavior as took place when the animals were pairing up, we can call it post-nuptial courtship.

By the very terms of our definition, mating behaviors serve a clear-cut, useful function in reproduction. They enable the male to inseminate the female. We can, therefore, readily understand them in terms of adaptive behavior. On the other hand, courting behaviors are often not only bizarre, wasteful, and useless, but, as with the snarling and "fighting" of cats, seem to interfere with the smooth progress of the mating process. Yet they are common and conspicuous parts of the reproductive behavior of many vertebrates. Why?

We will not attempt here to describe even a sampling of the great variety of courtship behavior to be found in nature. Though fascinating, this would be a confusing and endless task. Instead we will attempt to analyze the principal functions accomplished by courtship behaviors and illustrate these by appropriate examples.

FUNCTIONS OF COURTSHIP

Advertisement.—Clearly one of the problems in mating is the recognition of appropriate mates not only as to sex but also as to readiness to mate. Behaviors that "advertise" the sex and readiness of the mate constitute one class of courtship. Whereas the male of many vertebrate species is continuously sexually active during the reproductive season, the female is suitable as a mate only during a limited estrus. Therefore, the estrus female must provide the clues for mating to the male. Many of these are physiological rather than behavioral in nature. Secretions produced by the female in one fish were found to stimulate mating behavior in the male (Tavolga, 1956). Since most mammals are primarily nocturnal, non-visual animals, they use the sense of smell widely. Many female mammals in heat can apparently be recognized by odors, sometimes from special anal glands like those of the cat. Male rats distinguish estrus from non-estrus females, even at some distance, by odor, and male dogs are attracted to the urine of the estrus bitch. In female chimpanzees and some other female primates the estrus state includes swelling and reddening of the so-called sexual skin in the perineal region. In addition, the estrus female in mammals generally shows behavior that differs from the non-estrus condition. The non-estrus female is antagonistic to the approach of the male, avoiding or repelling his advances. As the minimum of estrus behavior then, the female modifies this repulsion so that she stands still for the male and tolerates his sniffing and eventually his mounting behavior. In many of the larger ungulates, such as deer, estrus behavior shows little more than this standing for the male. The female shrew will kill the male confined with her in a cage except when she is in estrus. Many rodents, rats, guinea pigs, etc., when stimulated by a male during estrus, show a special posture response called "lordosis." In this, the back is arched downward in the center, bringing the genital region up, and the tail, if present, is turned sharply to one side. This lordotic posture, of course, directly assists in permitting the male to achieve intromission. In fact, intromission cannot be achieved without such co-operation in these or in other mammals with the exception of man and perhaps some higher primates. In addition to lordosis, the female rat in heat shows a peculiar start-and-stop running when followed by a male. She is also characterized by a tremendous increase in running activity, which presumably facilitates contact with the male under natural conditions. The female lion not only exposes herself to the male but will directly crawl up to and squeeze underneath him if he is not otherwise responsive to her advances and odors. These advertisement displays of the female in estrus are, however, not a particularly elaborate aspect of courtship in vertebrates, because, as mentioned before, the male generally maintains sexual receptivity continuously during rut and actively seeks the female. The biological "need" for female "advertising" display is not great. Nevertheless, the female vertebrate generally does display a definite estrus period of receptivity even if, as in the guppy, it is so subtle that it is difficult for the human observer to identify it (Clark and Aronson, 1951).

Advertising displays of the male are a more conspicuous, though perhaps not more common, aspect of animal courtship. We appreciate this best in birds, lizards, and some fish which, like the human, are "visual-minded." The conspicuous "showing off" of fine feathers by the cock of the barnyard fowl and of many other birds is well known. Because in its extreme form this type of display seems more related to the correlation function of courtship than to simple advertisement, it will be discussed under this topic later. Here we may point to examples that more simply fit the concept of advertisement function. The male of the common heron of Europe builds a partial nest in the tree tops of the colony area. This is readily seen from above. He then

FIG. 4.8. In the Jackson Whydah, a bird of the plains of Kenya, many males establish their "courts" together. Each consists of a ring on the ground, about ten feet in radius, trodden flat around a central tuft. In this "court" the males advertise for females by fluttering up and down like "Yo-Yo toys." When a female lands in the arena the male raises and quivers his neck hackles and tail plumes and occasionally flounces his tail in the female's face. (After E. Armstrong, *Bird Display and Behavior*, 1947.)

places himself in this nest foundation and assumes a very awkward and conspicuous posture, erect with neck stretched, head pointing upward, and neck feathers fluffed out. At the same time, he calls with a loud sharp cry. This strained behavior may continue for many hours a day for many days until a female flying by is attracted and stays to mate (Stonor, 1940). For lek birds, such as the ruff and sage grouse (see Chap. 1), in which many males take up small mating territories close together, the essential function of the group display of fine feathers and noise would appear to be the advertisement of a group of males ready to mate (Fig. 4.8). Females in the estrus state can then readily ritory, he displays his red belly by turning vertically upward with his ventral side toward his opponent. A ripe female, on the other hand, turns up but reveals the egg-swollen silver belly instead. The territoral male's reaction to the male display is aggressive, but the display signals from the ripe female modify this aggression so that the territorial male turns and swims toward the nest. He turns again to attack, reverts again toward the nest, etc. This zigzag dance that thus characterizes the courtship is a modification of the aggressive behavior shown toward a male (Tinbergen, 1953).

The courtship behavior of the American song sparrow likewise is built around territorial aggression modified by the nature of the female reaction. If the male displays, sings, or even pounces on the female, instead of the female replying in kind as a male would, she persists in the territory without fleeing or showing aggression. Because of this receptive behavior, the male's display of aggression is soon diminished, and he becomes reconciled to her sharing his territory (Nice, 1937). This type of behavior is seen in many territorial birds. Even the heron, which goes to such lengths to display his availability as a mate, attacks the female with fierce determination when she finally does come to his nest. A primary factor in achieving the pairing is the ability of the female to accept the initial aggressiveness without fighting back. Since many birds are monomorphic, that is, male and female are alike, this subordination behavior seems to be the chief sign by which the sex of the individual is recognized. On the other hand, if the sexes differ morphologically, this serves as a means of sex identification. The male flicker has a mustache-like arrangement of black feathers lacking in the female. After some black feathers were attached to a trapped female to form a "mustache," she was attacked by her mate when she was permitted to fly back to her nest (Noble, 1945).

Physiological co-ordination of male and female.—Successful reproduction requires as a minimum the co-ordination of male and female activity with respect to insemination. In many vertebrates, most particularly in birds, the co-ordination must extend far beyond that of insemination. In some, both parents co-operate in a long process beginning with territory establishment, continuing through nest building, copulation, egg laying, incubation, brooding, and feeding of the young until they are ready to leave the nest. It is not surprising, therefore, to find in such birds courtship activities that are not only elaborate but extend through the entire reproductive season. Sho

find suitable mates (Armstrong, 1947). The male of many territorial mammals, in marking off his territory (by odors as the dog does with his urination stations or by sound as the moose does by bugling), not only notifies other males of his territorial rights but presumably communicates relevant information to the females. The loud singing of territorial passerine birds likewise serves to alert females to an available territory. The posing and flashing of his brightly colored gularskin flap by the male anolis lizard in his territory has been found to attract females from afar (Greenberg and Noble, 1944). Similarly, the patrolling of territory boundaries by conspicuously marked males in fish such as the sticklebacks also serves an advertising function.

Overcoming of aggression.—On the whole, the advertising function of courtship accounts for only the simpler aspects of courtship. Another function which courtship serves is that of overcoming the aggressive responses of one animal to another. Such aggression may be considered here under these three headings: the aggression of predators, of dominant animals, and of territory owner.

(*a*) Predators: The behavioral mechanisms of a predaceous animal, of course, prompt it to attack any other animal approaching it. It is clear that if male and female are to mate successfully, the predaceous behaviors of the mates must be overcome by some signalling system. This task is often accomplished by conspicuous courtship behaviors. Perhaps the classic example of this is seen in various species of jumping spiders, which feed by pouncing upon their prey. Since the female is larger than the male, any male approaching a female is in danger of ending up as food rather than as mate. In such species the males approach the females with elaborate courtship dances. While still a safe distance away, the male assumes unusual postures rising high on its legs and waving conspicuously marked appendages before the female. Such dancing may last for hours, and mating is not consummated unless the female shows appropriate signs of quiescence (Fig. 4.9). Some predaceous flies wrap a victim in silk and present this to the female as part of the courtship; and, according to the usual interpretation, while she is occupied with this "present," mating is successfully accomplished. Still more bizarre actions have been observed, such as the female spider going into catalepsy before the male makes his final approach or submitting to being tied firmly to the ground with silk spun by the male (Bristowe, 1941).

Solitary predaceous mammals, particularly of the cat and weasel family, often do considerable fighting as part of their courtship. The

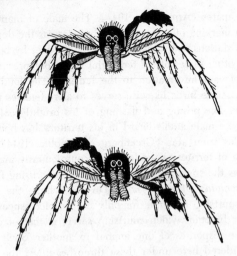

FIG. 4.9. The male of some spiders courts the female by waving his conspicuously banded appendages. The female is induced to go into a trance-like state during which the male safely mates with her. (Redrawn after Bristowe, 1941, and M. Burton, *Animal Courtship*, 1953.)

mating of cats is accompanied by much scratching, biting, and appropriate vocalization. This behavior differs from true fighting, for only minor injuries are inflicted; and since neither animal yields, it continues for some time. The mating process of ferrets and mink looks like a prolonged fight, often lasting an hour or more. The courtship fight arouses the animals to a high level of emotional tension which, as we shall see later, has important physiological consequences. Behaviorally, it appears to be fighting behavior normal to a predator but modified and controlled by stimuli emanating from the sexual situation.

(*b*) Dominants: The aggressive behavior of group-living animals is, as we have seen in Chapter 1, subjected to patterned control by the formation of dominance hierarchies. The development of mating behavior between males and females often involves important modification in these dominance-subordination patterns. In groups such as those of the baboon or the Indian antelope where the male overlord is supremely dominant, the other members of the group pattern their movements in relation to his, keeping out of his way, permitting him to take over any desirable object without dispute and yet remaining near —but not too near— him. When a female baboon comes into heat, how-

ever, she comes close to the male, standing her ground when he makes aggressive passes at her in a way which she never does when not in heat. She also displays her genitals in the so-called presentation posture. In this pose the animals faces away from the dominant, bends over at the waist in such a way as to expose the perineal region to the view of the male, meanwhile watching him with sharply turned head. At this time her sexual skin is engorged and highly colored, revealing her estrus state. This posture affords maximal convenience for the male to mount her (Zuckerman, 1932). Presentation is a common sign of submission to a dominant in many primates and is adopted by male subordinates as well as female. Thus courtship behavior in the female monkey consists principally of subordination, sexual receptivity, and acceptance of punishment from the dominant male. Male dominance is mitigated to some extent, since the punishment administered to the estrus female when she violates his dominance privacy is relatively mild and soon stops altogether (Carpenter, 1942). It has been shown experimentally that the female chimpanzee becomes more assertive toward her mate, often assuming a typically dominant acquisitiveness during estrus. (Birch and Clark, 1946) Alteration in dominance at estrus has also been recorded for many animals. In some birds, such as the budgerigar or parakeet, the female generally dominates the male except when she enters estrus. A female pigeon has been observed to rise in dominance position with respect to other females upon pairing with a dominant male, and similar interactions of dominance and mating associations has been reported in the Japanese monkey (Imanishi, 1957).

(*c*) Territory Owners: One of the commonest forms of aggressive behavior, as we have seen in Chapter 1, is that connected with territoriality. The common form of individual territory in vertebrates is that set up and defended by the male. Obviously, if mating is to be achieved in such forms, the female must somehow gain access to the territory and achieve a *modus vivendi* with the male. The modifications in behavior which are involved and which we here consider to be courtship may be centered around three main techniques: (1) display of morphological sex recognition marks, (2) behavioral characteristics marking the female as distinct from the male, and (3) display of subordination behavior.

When a stickleback swims into a male's territory, the owner swims to attack. If the other animal is a male or an unripe female, it generally swims away. If the invading male is "inclined" to dispute the te

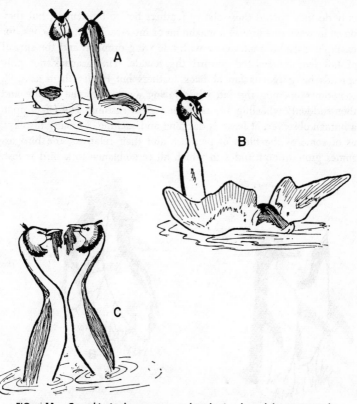

FIG. 4.11.   Courtship in the great crested grebe involves elaborate mutual ceremonials. Some incidents of these activities are illustrated above. At A we see the head shaking ceremony in which the pair face each other displaying their head feathers and shaking their heads from side to side. B shows the male dive in which the male approaches the female with head submerged and suddenly shoots high out of the water just in front of her. At C we see the mutual presentation of water weeds to each other. (After J. S. Huxley, Proc. Zool. Soc., 1914.)

tinguish male from female without behavioral signs in gulls, song sparrows, etc. Some of the most striking, and certainly the most conspicuous, types of courtship occur in those animals in which male and female are dissimilar, i.e., dimorphic species. In these species, it is almost always the male that is the brightly plumaged sex. The barnyard rooster, the turkey cock, the bird of paradise, and the peacock are well-known examples. Pictures of their fancy feathers, however, give only a faint idea of the true nature of their courtship. Not only do they spread these showy feathers before the female, but they do so in ways that arouse a maximum of interest. The turkey cock, for example, parades a side view with one wing dropped and the spread of tail feathers twisted toward the female. The peacock not only spreads his gorgeous fan of back feathers but rustles them as well, sometimes opening the fan while facing away from the female and then suddenly wheeling around with a great rustling. The effect upon a human observer, at least, is dazzling and exciting. In many displays, as of some of the birds of paradise and their relatives, the bird assumes grotesque attitudes in which all resemblance to a bird is lost.

FIG. 4.12. The yellow-eyed penguin often courts the female with the posture called the "salute" (A). This same posture is often shown as a social greeting between any two animals and is not restricted to courtship activities although it appears to be common then. The "ecstatic" posture of the erect crested penguin is a mutual ceremony in which a pair indulge during pair formation and afterward during post-nuptial courtship, when one individual relieves the other in incubating the egg. First the birds pose upright (B), then bow deeply, rocking from side to side and calling loudly (C). (After L. E. Richdale, *Sexual Behavior in Penguins*, 1951.)

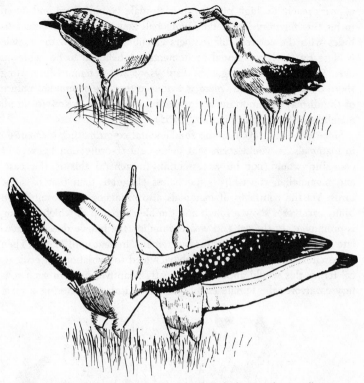

FIG. 4.13. The courtship dance activities of the wandering albatross begin with mutual nibbling of the beaks (*above*) and reaches a climax with the ecstatic display of the huge wings by both male and female. (Drawn from a photograph.)

If sounds accompany the display, they are loud and raucous rather than melodious. All in all, such displays clearly tend to arouse a high pitch of excitement in the female. The functional significance of these displays must lie in the stimulation of the female to the mating act. As we shall see in the discussion of the physiological correlates of these phenomena, there is much evidence that such courting excitements stimulate the reproductive mechanisms of the female.

The effect of a single displaying animal is exaggerated when the display is part of a communal activity (Fig. 4.14). Even in non-colonial birds, sexual excitement is contagious (Collias and Jahn, 1959). In fish the turmoil of breeding aggregations seems to have stimulatory value (Aronson, 1957). The excitement at a ruff lek when a female appears reaches a high pitch with each male bounding up and down in his tiny territory like a Ping-pong ball and assuming ecstatic attitudes with the colored ruff feathers extended fully when the female is close by. Such communal excitements are believed to be achieved even in the much less garish displays of the communally nesting shore birds. In the gulls, there is evidence that the communal nature of the display is important in correlating the phases of activity in all members of the colony (Darling, 1952).

Since the estrus cycle is the fundamental co-ordinating mechanism in mammals, it would seem that behavioral co-ordination by way of courtship would not be as essential. In general this is the case, and mammalian courtship is much less elaborate than that of some birds. Yet the courtship of mammals also produces great excitement. Many carnivores show a rough and tumble courtship in which fighting is prominent. In the cat and weasel families, the female does not ovulate spontaneously during estrus but remains in heat until mated. Then the excitements of the mating behavior lead to ovulation. In ecological terms, this would appear to be neatly adaptive, since in such solitary carnivores the female cannot be assured of encountering a male

FIG. 4.14. The prairie hen males gather in a lek and display by inflating the large yellow air sacs on the neck and bowing with tail fan spread when the females approach. (After W. P. Pycraft, *Courtship of Animals*, 1914.)

as soon as she reaches the proper stage of estrus, in spite of powerful odor and seeking behavior. The natural pugnacity of the carnivore has thus been incorporated into the courtship of the animal and aids reproductive co-ordination by prompting ovulation.

The domestic rabbit is also known to remain in estrus without ovulation until released by the excitements of mating, although here mating is not accompanied by such rough-house play. This might be deemed a contradiction to the hypothesis formulated above, but perhaps it too can be understood in ecological terms. The female of the European wild rabbit is larger than the male and maintains a defended territory around her warren from which she excludes the smaller male. As might be expected under these circumstances, the male has perfected a courtship that is appropriately cautious but exciting to the female. He may walk in a stiff, high-stepping manner away from the female or around her, in either case with tail raised and rear quarters directed toward the female. Once the male excites and possibly dominates the female, he directs a jet of urine at her. Such "enurination" behavior is occasionally also seen in states of excitement provoked by being trapped, for instance. It is possible that the dependence of ovulation in the rabbit is a reflection of the difficulty in mating that results from the necessity of overcoming the territorial defense of the female. At any rate, in these few mammals, coordination must be regarded as a function of their courtship activities.

Display as a General Phenomenon

If we think back at what has been said about display in relation to territorial defense, dominance patterning, and courtship, we recognize that these involve ostentatious "attention-calling" behaviors. All of these are of course related to reproduction to some extent; yet it is important to note that the functions of display extend beyond the strictly sexual relations of mates. In territorial display, the showiness is often directed not to the mate but to sexual rivals. This factor accounts for the prominent part played by aggression in the expressions of some displays. Because of the close interlocking of territoriality, dominance, and the reproductive activities, we cannot really identify many of these displays as being properly in one category or another; that is, reproductive behavior in social animals tends to become intermixed with these other two behavioral types so that characters which seem appropriate for one activity become deeply involved in the others. Thus defense of territory, rather than mating, seems to be the goal of "sexual" behavior in many cases. Some species of seals, for example, pay no attention to the females which may leave a territory; the male makes little attempt to herd them in yet he defends his territory zealously. In some gulls the attachment of the parent seems to be primarily to the territory, since this is ardently defended; whereas the eggs, nest, or mate can be changed around at certain stages without arousing any compensating behavior. From a behavioral standpoint, therefore, the sexual drive of an animal may be expressed more strongly in "irrelevant" activities than in those directly related to reproduction. It cannot, therefore, be properly said that defense of territory is a displacement of sexual activity or that dominance fighting by males is in any sense misdirected sexual motivation. According to the nature of the species, these activities may be of the essence of the expression of sexual motivation.

Parental Care

Without attempting a systematic discussion of parental care in the animal world, we may consider here certain aspects which were not conveniently discussed earlier and which illustrate the adaptive nature of the variations to be found.

PARENT-YOUNG INTERCHANGE

Among colonial insects, such as ants, termites, bees, and some wasps, it has been found that members of the colony actively exchange food by regurgitation. If dyed-sugar water is fed to some members of an ant colony, the color is seen throughout the colony within a few days. This exchange of food is called "trophallaxis." It plays an important part in the correlation of the activities of the insect colony. In termites, for example, it is believed that substances transmitted in the trophallactic pool determine the differentiation pattern of the young with respect to caste specialization. This trophallactic exchange also plays a role in parent-young relations (Michener and Michener, 1951). In ants, for example, the larvae produce fatty secretions for which the adults have a great appetite. The feeding of young by adults is thus an exchange phenomenon dependent upon the specific appetites, not upon young-parent recognition. This is illustrated by the success certain nest parasites, such as the Lomochusa beetle, achieve because they produce the sought-after secretions in even higher quantity than ant larvae. The adult ants feed them in preference to, and to the detriment of, their own larvae. The concept of a mutual exchange forming the behavioral basis for parent-young interrelations may be applied to many activities other than food exchange. In army ants, for example, the activities of the young in various stages of their development appear to provide the stimuli that regulate colony activities (Schneirla, 1938). Whether or not we use the word trophallaxis to describe the exchanges provided by stimuli other than food, the basic idea that the relation between parent and young is integrated by a mutual exchange rather than by a one-way giving is applicable to many other organisms.

An almost universal characteristic of mammalian maternal behavior is the eating of the placenta and the licking of the young by the mother. Even guinea pigs and many herbivores which normally do not eat meat at all eagerly consume the afterbirth. The licking of the young has certain obvious practical values—it cleans and dries them and clears their nose, mouth, eyes, and ears. The eating of the placenta, despite its high hormone content, is not definitely known to be of any practical importance for the mother. Only the higher primates as a group seem to lack this licking and eating behavior, although among domesticated herbivores (cows, for instance), the failure to eat the placenta is not uncommon, and there are several species of mammals, such as the wild pig and fur seal, which apparently do not eat the afterbirth (Hediger, 1955).

From the behavioral standpoint, the post-parturitional licking activities of the mother appear to be important in establishing the mother-young bond. It has been found in goats that if the young is removed immediately after birth and before the mother has licked it and is cleaned up and returned a few hours later to the mother, she will refuse to accept it. She no longer has the drive to lick the baby. As the young approaches and tries to nurse, she drives it away just as a normal goat drives away foreign young that try to nurse.

The ability of the mother to become associated with her own young seems to be limited to a short period, two or three hours in goats and sheep. Even slight interferences with the normal relationship during this critical time has far-reaching effects on the mother-young relationship. Experimental disruptions at this time, even if they do not prevent the baby from nursing but merely shorten the licking process, may result in a failure in the completeness of the mother's response to the young when they call from a distance (Blauvelt, 1955). The cleaning of the new-born chimpanzee is accomplished by finger manipulation by the mother. The behavior is akin to grooming and is visually guided and not dependent upon the chemical senses. As we shall see later, it also does not have as firm an instinctual basis as the licking behavior appears to have. In herd-living mammals the initial relationship between mother and offspring seems to be that of the mother recognizing her own young. The young soon learn to associate specifically with their own mothers presumably as a result of rejection by all other females. Thus, herds of many ungulates such as mountain sheep and deer have as their primary group the maternal family in which the young are associated with their mothers and these, in turn, with their own mothers as long as the latter survive. Since the mother retains a dominance and leadership relation to her young, the herd constitutes primarily a matriarchy with one or a few oldest females exerting social control (Darling, 1937; Murie, 1944).

A phenomenon analogous in many respects to what we have described above for mammals seems to take place in some birds during the establishment of parent-young relationship. Precocial birds, which follow their parent soon after hatching, have been found to become "imprinted" with the characteristics of the parent or surrogate parent during the first few hours after hatching. If a turkey or greylag goose is incubator-hatched and cared for by a human in its initial hours, it will thereafter show no fear of human beings but will instead react to them as it normally would to a parent. Greylag geese thus imprinted follow a person and will not follow adult geese. After the first few hours, the hatchlings lose this capacity for imprinting and will not form the association. Once imprinting is accomplished, it cannot be displaced from the imprinted object by other associations after this time. It should be noted that the process of imprinting is a young-to-parent relation, not a parent-to-young relation as we saw it develop in the ewe. The imprinting process shows a surprisingly long-range effect. For example, hatchling jackdaws imprinted to humans displayed courtship behavior only to humans and not to fellow jackdaws when they become sexually mature many months later (Lorenz, 1952). In birds that are hatched in a very immature condition and remain in the nest for a long period (altricial birds), the identification of young to parent, in so far as it occurs at all, is a more gradual process and takes place later.

The parent-young relationship is the primary social relationship in birds and mammals generally. It may remain in force with modifications throughout the life of the animal, as we mentioned above in the case of the red deer, and constitute the basis for herd organization. In mammals which live in permanent groups, this mother-young relationship is the basic factor in social organization. In birds the parent-young relationship usually disintegrates before the coming of winter. The migratory flocks which form then do not have a familial basis. Often older males, older females, and yearlings form separate flocks for migration.

In both birds and mammals the breakup of the original family relationship has been ascribed, at least in some instances, to positive disruptive behaviors rather than simple cessation of the bonds that held the individuals together. Thus when the female of many species of deer arrives at late pregnancy, she adopts a negative attitude to her yearlings and when they attempt to nurse rejects them with threats, butts, or even biting (Altmann, 1952). Mother bears actively take steps to break the associations between themselves and their cubs. In chickens, too, an active driving-off of the young is seen when the hen is ready to enter a new laying period. Where there are a few dominant males in the social groups, as in macaque monkeys, these males drive out the young males as they mature but do not eject the females. As a consequence, these monkey colonies generally have small bachelor groups which associate peripherally with them. The young males in this group are ever alert for an opportunity to invade the group in order to achieve a permanent place and gain access to the females.

TEMPORAL LIMITATIONS

Another aspect of parental behavior that deserves particular comment is the timed, sequential co-ordination of the separate events. The fanning activity of the male stickleback, for example, is part of a timed, sequential behavior pattern. It begins only after a female has laid eggs in the nest and increases in intensity gradually and appropriately as the young develop. When they emerge from the nest, the fanning ceases abruptly. Similarly, the retrieving of young which have been removed from the nest constitutes one of the characteristic parental behaviors of mammals such as the rat. This behavioral response to misplaced young appears characteristically in the female late in pregnancy. It reaches a high peak during the first days after parturition and declines gradually so that by the time the young are ready for weaning on the twentieth day or so, retrieving behavior has practically disappeared. But, of course, it is in birds with their highly complex parental behaviors that we find this timing factor most prominent. A definite migration-restlessness appears in the appropriate season; territory-establishing behavior follows and lasts only a limited time. Song birds develop an interest in nest sites and nesting behavior which ceases at an appropriate time. Only then do the mates commonly show coitional behavior. This in turn may last only a few days and then disappear entirely from the repertory of the pair, and so on, through the entire sequence of their reproductive behaviors (Lehrman, 1961). We have previously emphasized the importance of courtship, territoriality, flock organization, and parent-young relations in maintaining the behavioral contacts and associations that permit necessary co-ordination. It must, however, be emphasized that the problems presented by the intricacy of these co-ordinations are among the most baffling in behavior study (Marshall, 1961).

Sexual Dimorphism as a General Phenomenon

The sexual structures of animals are generally classified as primary and secondary. The primary include the sex glands, their ducts, and other accessories which serve the basic reproductive mechanisms characteristic of the taxonomic group. Since reproductive processes are fundamentally the same in all members of a group such as birds or placental mammals, the primary sex characteristics do not differ greatly between species in such groups. The secondary characteristics consist of those relatively superficial characteristics such as display feathers, horns, etc., which distinguish male from female of a species but which do not play a direct role in reproductive physiology. These are found to vary greatly from species to species and to be correlated with the detailed behavioral characteristics of the species. Sometimes another term, "epigamic," is used to refer to characteristics which, like secondary sex characteristics, play an indirect role in reproduction but which, unlike secondary sex characteristics, do not necessarily differ between male and female. The plumes of night herons or the red breast of robins might be cited as examples. These are alike in male and female but play a role in display behavior during reproduction. From the point of view of behavior, the important concept to be developed is that the secondary sex, or the epigamic, characteristics are closely related to behavior and often hold the clue to an understanding of the behavioral organization of a species.

One type of secondary sex characteristic may be described as physiological. It consists of a sensory-motor differentiation based on sex. This type is common among insects. The glow worm, for example, is the wingless, grub-like female of a firefly beetle. In many insects the male is differentiated as the motile, active, seeking type, whereas the female is the larger, food-storing specialist. This is analogous to the primary differentiation of the sperm and egg. This type of differentiation is readily understandable on physiological grounds since it contributes to the efficiency of the reproductive process. We would,

FIG. 4.15. The male (right) in the American bison illustrates the common characteristics of dimorphism in social-living mammals. The male is larger, especially in the fore-quarters. Head and horns are enlarged as part of the emphasis on aggressive potential. (After W. P. Pycraft, Courtship of Animals, 1914.)

therefore, expect to find it widespread. Yet it is most significant to note that among vertebrates this type of differentiation between the sexes is actually a rarity. An example is the deep-sea angler fish in which the female is large and fully developed, whereas the male is reduced to a tiny parasite.

A second type of secondary sex characteristic, which may be designated as aggressive potential, is the difference between male and female with regard to capacity for fighting (Fig. 4.15). This type is common among vertebrates. Most prominent among these dimorphisms are differences in size and strength. One of the extreme examples is seen in the seals and related marine carnivores. Elephant-seal males are as much as two and a half, and fur-seal males ten, times as large as their females (Bartholomew, 1952). Though this is extreme, a difference of 50 per cent or so is not at all rare among mammals.

In a majority of mammals, the male tends to be bigger and heavier than the female. Only exceptionally, as in the European rabbit, is the female the larger.

Aggressive potential in favor of the male often takes the form of weapons. Horns and antlers are in many instances differentiated between sexes. We are familiar with them in many species of deer. Teeth as weapons are also frequent secondary sex characteristics of mammalian males. We see this in the enlarged canine teeth in male baboons and, in extreme form, in the single large tooth of the narwal. In birds, examples of dimorphism in weapons are fewer, but the spurs of the rooster provide a good one.

A third category of secondary sex characteristics may be designated display characteristics. We have discussed many examples of these

FIG. 4.16. In the courtship of sexually dimorphic animals, the conspicuously colored male generally deports himself in such a way as to display his plumage patterns before the female. The snow bunting male, for example, faces away from the female and spreads his feathers stiffly, thereby exposing the conspicuous patterns of his back. He then runs quickly away from her for a short distance and repeats this performance again and again. (From N. Tinbergen, *Trans. Linn. Soc.*, 5 [1939].)

in the conspicuous plumage of many male birds (Fig. 4.16) and the color display of sexual skin in female primates. Perhaps some fancy furs among mammals deserve special mention. The male of many deer develops a thick coat of long and light-colored fur over the neck at rutting season. The lateral presentation of the "fur collar" is a prominent feature of display. The mane of the lion is probably to be regarded in the same category. Most of the better-known examples of sexual dimorphism among lower vertebrates also fall into this category. The brilliant colors of lizards and various aquarium fish, such as the fighting betta, immediately come to mind.

Several important points emerge from this brief consideration of secondary sex characteristics. It might be expected that natural selection would favor the physiological type of dimorphism, since each sex is thereby better equipped for its natural role. The female would be expected to be larger, since she must furnish the egg with its food supply or protect and nourish the young. The males would be expected to be, like sperm, small, numerous, and equipped with sensory-motor mechanisms that enabled them to seek out the larger and more sluggish females. As we have noted, such dimorphism is commonly found among invertebrates, particularly insects. Vertebrates, on the other hand, emphasize dimorphism of aggressive and display potential. In some ways this is the direct opposite to what might be expected.

The explanation of this paradox is to be found in the peculiarities of the position of the male in vertebrate reproductive patterns. Often the male contributes little more than the sperm transmitted to the female in the act of insemination. Yet to achieve insemination, he must compete with other males. Furthermore, since one male is physiologically capable of inseminating many females, a high survival rate for the males is of small consequence to the survival of the species. The selective processes to which such males are exposed thus favor characteristics which give success in competition with other males, with little regard for the deleterious effects which such characteristics have on the survival of the males themselves. Selection in females, on the other hand, is much more conservatively managed, for a high survival rate of the female is of primary importance, since care and provision for the young are dependent upon the female. Her more limited reproductive potential, moreover, makes her survival of great importance for the species' success. In this sense the males may be described as expendable in contrast to the females.

We can thus visualize that in the typical herd-living mammal, such as the deer, the primary determinant of biological success in a male is the capacity for maintaining a large harem in competition with other males. Characteristics of aggressive potential and display have been pushed by selection to the point of decreasing the chance of ultimate survival of the males bearing them. Of course, the need for individual survival cannot be entirely ignored, and there are limits beyond which the unphysiological specialization of the male for aggression or display cannot go. But it is clear that a characteristic that insures reproductive success may be favored even at the expense of individual survival. This is especially clearly shown in polygynous mammals. The existence of these selection pressures enables us to understand the general unphysiological specialization of males in many vertebrates and the well-known fact that, from a physiological view as measured by life span, resistance to disease, etc., the male is commonly inferior to the female of the same species.

It can be appreciated that the above reasoning applies particularly to mammals which live in groups and in which the male plays no role in parental care. It also applies to birds and other vertebrates under the same limitations. If we examine the instances of high development of display potential in birds, we see that the same principle can be invoked here. The birds which show the most elaborate development of conspicuous feathers are birds in which the male plays little or no role in caring for the young. The gallinaceous birds, lek-breeding birds, birds of paradise, etc., are all birds in which the parental functions are performed by the female to the practical exclusion of the male.

On the other hand, if we consider vertebrates in which the role of the male and female in care of the young is more nearly equal, such as the shore birds and most song birds, we find that the species is generally monomorphic, i.e., males and females are alike (Fig. 4.17). In these forms, we may find a moderate development of display potential in epigamic characteristics. This has its explanation in the value of the displays for mutual courtship and for territory defense on the part of both male and female. We do not, however, see such display potential carried to the point of interfering with individual survival by unduly exposing the bearers to predation.

FIG. 4.17. One form of nest-relief ceremony in the herring gull is for the bird returning to the nest to bring nesting material, as in courtship. Both parents in this monomorphic species share in care of the young. (After N. Tinbergen, *Herring Gull's World*, 1953.)

In exceptional instances, as in button quail and the phalaropes, where the behavioral role of the male and the female are reversed, it is not surprising to find that dimorphism also assumes a reverse pattern (Kendeigh, 1952). The female here is the showy-feathered member of the species (Fig. 4.18). Of course, in the vast realm of nature, where many other factors come into play, some exceptions occur, but it is an important general principle that the type of sexual dimorphism shown by a species correlates with the role of the sexes

FIG. 4.18. The male of the red-necked phalarope is smaller and less brightly plumaged than the female; he has considerably less red coloring on his throat, and the white stripe over the eye is inconspicuous. The female courts the male by rising up in the water, fluttering the wings, and calling loudly. Mating takes place in the water. The male builds the nest on the shore, and after the female lays the eggs, he assumes the entire burden of incubating. He pulls the tall grass over himself after settling on the nest for concealment. (Drawn from a photograph.)

in courtship and parental activities. In Chapter 4 we shall see the applicability of this principle to the understanding of social organization among vertebrates.

WILLIAM ETKIN

3

Theories of Socialization and Communication

Introduction

Our previous chapters have given us some insight into the ways in which social animal groups are organized, particularly in so far as that organization is determined by social status, territoriality, and reproductive activities. Except incidentally, they have not, however, enabled us to study the attractions which keep animals together in groups. We now wish to consider the concepts zoologists have developed about the nature of these attractive forces. For that purpose, we shall inquire into current ideas of the role, which innately determined behaviors and learning or experience play in the bonds that hold animals in groups.

For many years the concept of instinct as part of the behavioral equipment of animals has been in disfavor with many American psychologists. We need not go into the many reasons for this nor attempt to evaluate the position of those psychologists who still regard the concept as undesirable. It will be sufficient to say that, in this author's opinion, the concept has acquired new and significant meaning in the study of animal behavior, and its elaboration has provided insights into the nature of social behavior that are indispensable for modern views.

The present-day rehabilitation of the concept has resulted in large part from the penetrating observational and experimental work of some European zoologists, particularly Konrad Lorenz and Niko Tinbergen. In their writings these authors have expounded the value of comparative study of animal behavior under natural conditions or of laboratory study which utilizes methods and problems suggested by field observations. The name "ethology" has been given to this field to distinguish it from the more conventional animal laboratory studies of problems many of which have been suggested by human psychology. Our concern will be with only those aspects of ethological theory which help us to understand the social behavior of animals. We shall attempt to present this theory in its first, or classical, form as developed since 1930 (for summary, see Tinbergen, 1951; Lorenz, 1957) and shall refer only briefly to subsequent modifications by various ethologists. One reason for this procedure is that an understanding of present-day developments requires knowledge of the original theory. A second reason is that, as in so many other areas, research has gone off in many directions without producing a clear consensus of new understandings to replace those found wanting in the older theory.

Instincts will be viewed here as complex behaviors which function as adaptive units of action in an animal's life and which appear to be largely independent of learning. In other words, when we say of a behavioral pattern that it is an instinct, we assert that we believe that its fundamental elements are innately organized and require no specific type of experience as background for effective performance. Of course, the term has been used in other meanings in various philosophic interpretations of behavior, but we are here concerned only with the meaning that is operationally useful in experimental analysis. There is no a priori reason to expect that any activities as complex as the common instincts should necessarily be entirely uninfluenced by the previous experience of the animal. We find, in fact, that the interplay of experiential and innate factors in the development of behavior is very subtle and difficult to disentangle, a point repeatedly in evidence in the experimental chapters of this book.

By an instinct we also mean something that operates in the normal life of an animal as a means of accomplishing one of the activities required for its survival. In this respect, it is not a reflex, for a reflex is a bit of behavior that is not "goal-directed," in an adaptive sense, though it may, of course, be part of such behavior. Thus the withdrawal of a limb is a reflex which may be conditioned or unconditioned, but the setting up and defense of a territory appears to be an instinct, since it is a complex behavior functioning as a unit in the animal's life history and apparently not requiring specific experience with territories for its adequate expression in the adult.

The concept of "innateness" has been subjected to vigorous criticism by many workers, particularly Lehrman (1953) and Schneirla (1956). Without attempting to analyze this controversial facet of psychological theory, it would perhaps be well to clarify the use of terms in the present chapter. By "innate," in reference to either a structure or a behavior, we mean so regulated by hereditary factors as to require no specific conditions external to the organism for its development. Thus we say that eye color is innately determined in *Drosophila* and man. We are, of course, aware that there are an enormous number of interacting activities that intervene between the gene and its expression in the animal. These factors include the interrelations of various parts of the organism, particularly inductive, hormonal, and neurotrophic interactions. The general conditions of the environment also have to be favorable. In this sense, the eye primordia must have certain interactions with other parts of the organism and with generalized (nutritional, etc.) factors from the environment in order to develop successfully and express their "innate" characteristics. The word "innate," therefore, refers to a complex idea, not to a unitary factor or one that is entirely autonomous (cf. Lehrman, 1953, p. 343). For example, some cancers in mice are thought to be determined by innate factors, others are said to be acquired. In the former no special external condition is required; in the latter exposure to a specific virus or other agent is a necessary antecedent condition. Similarly, we shall speak of behaviors like leg-lifting by male dogs as innately determined. We imply by this that the place to look for the influencing factors is in the developmental physiology of the animal, including hormonal changes, etc. On the other hand, we think of retrieving behavior in the dog as learned behavior, and so if we wish to understand this behavior in a particular dog, we seek information on the special training that dog received prior to performing this particular act. In either case we could be mistaken about our assumptions, but the initial characterization is a useful first step, based on our unanalyzed experience with the behaviors in question.

The Lehrman-Schneirla criticism has served at least one very useful function. It has shown clearly that the adjective "innate" does not constitute an explanation. This criticism has become the focus of experimental analyses that have successfully shown the complex interaction of factors necessary for the development of certain behaviors. In showing the complexity of developmental and experiential interaction involved in behaviors formerly thought to be the simple and inevitable outcomes of genetic action, these experimental analyses support the contention of Lehrman and Schneirla—that the term "innate" is too vague to be useful. One of the valuable perspectives of ethology, however, is its view of behavior as part of the huge complex of animal adaptations to the environment. In this context "innate" seems very useful to the present author as a term to distinguish between those behaviors which seem to require and those which do not seem to require specific (learning) experiences for their expression. In that provisional sense we shall distinguish behaviors as innate or acquired. For example, we want to know whether the social behaviors described in earlier chapters, such as the selection and response to the appropriate mate in reproduction, require a special type of social experience on the part of the animal or can develop without any such experience. At least at the level of analysis of this chapter, the distinction between innate, instinctual patterns and acquired, or learned, patterns is useful. It is perhaps pertinent to remark that the difficulty in dealing with the concept of "innate" behaviors comes partly from the vagueness of the antithetic concept, "learned" behaviors, since not all alterations of behavior resulting from experience fit our concept of learning. For example, the strengthening of muscles with exercise or the increased response of an organ to a hormone or stimulus after previous exposure without reinforcement would not seem to constitute learning. The concept of learning is as difficult to delimit as that of instinct. In consequence, there has been some tendency to substitute the term "experiential factor" for that of "learning" and thus avoid rigidity in thinking about specific problems.

The Ethological Concept of Instinct

THE MOTOR COMPONENT

We may analyze an instinct in terms of the stimuli which evoke it and the nature of the responses to those stimuli. It will be convenient to consider first the responses, that is, the motor component. These are complex actions, such as pecking at particles on the ground in chicks or mating performance in rats. Since these tend to be highly uniform and recognizable as units, they have been called "fixed-action patterns." We have described some of these in Chapters 1 and 2.

According to ethological theory as originally developed by Lorenz and Tinbergen, the co-ordination of motor activities is accomplished in an organized part of the central nervous system called the "instinct center." It is important to note that such a center is not necessarily localized in one group of neurones or even in one region of the central nervous system. The concept is a physiological, not a morphological, one. Its important characteristic is that the parts are so linked together that once activated they provide for that appropriate sequence of muscular or glandular activities which constitute the behavior.

Such a concept of a center of organization is, of course, not particularly novel. The important aspect of ethological theory are the properties ascribed to the system. One of these is that in such a center, a specific type of energy is generated which tends to produce a pressure for discharge and without which the center does not discharge. The basis for this idea of action-specific energy are experiments indicating that the threshold for the response to a given stimulus is increased after draining of the energy by discharge of a center and is lowered by a period of rest which enables the energy to reaccumulate. This is a very generally observed phenomenon. It is seen, for example, in the decreased response of a rat to a female after copulation, the decreased hiding response of a bird to the stimulus of a predator, the cessation of nest building in a stickleback after a bout of nest building, or the failure of the flying-up response of a male grayling butterfly when repeatedly stimulated by models of a female (Tinbergen, 1951; Lorenz, 1957). Such decreases in vigor of responses can be shown not to depend upon fatigue of the motor organs, because these organs can be made to respond in connection with other behaviors without loss of effectiveness. Similarly, the sense organs can be shown not to be fatigued. Therefore, the change must be ascribed to some loss in the central nervous system. This loss is limited to the action in question, to a reduction of "drive" in that system. Therefore, it is inferred that the center for the organization of each particular type of instinctual behavior accumulates a limited supply of energy. This is depleted by repeated activation of the system. When none is left, the system cannot activate its characteristic behavior in spite of appropriate stimulation.

Another source of evidence for the concept of action-specific energy is the existence of so-called vacuum activity, that is, discharge of an instinctual action in the absence of any apparent stimulus. Lorenz (1957, p. 143) describes an instance of such behavior as follows:

I had once reared a young starling who performed the whole behavior patterns of a flyhunt from a vantage point *in vacuo*, with a wealth of detail that even I had, until then, regarded as purposive rather than instinctive. The starling flew up onto the head of a bronze statue in our living room and steadily searched the "sky" for flying insects, although there were none on the ceiling. Suddenly its whole behavior showed that it had sighted a flying prey. With head and eyes the bird made a motion as though following a flying insect with its gaze; its posture tautened; it took off, snapped, returned to its perch, and with its bill performed the sideways lashing, tossing motions with which many insectivorous birds slay their prey against whatever they happen to be sitting upon. Then the starling swallowed several times, whereupon its closely laid plumage loosened up somewhat, and there often ensued a quivering reflex, exactly as it does after real satiation. The bird's entire behavior, especially just before it took off, was so convincing, so deceptively like a normal process with survival value, that I climbed a chair not once, but many times, to check if some tiny insect had not after all escaped me. But there really were none.

Another important behavioral characteristic that may be understood in terms of action-specific energy is that of displacement activity. It appears that when an instinctual activity is aroused strongly but its expression is at the same time inhibited either by simultaneous activation of another and incompatible instinct or by failure of appropriate releasing stimuli to appear, the animal may exhibit a different behavior pattern and one irrelevant to the situation. For example, fighting roosters when evenly matched often show evidence that fear is inhibiting their aggressive tendencies; the incompatible responses of flight and fight are both aroused. In such a case the animals will often break off their fighting and peck at the ground as though suddenly interested in feeding on non-existent grain. According to the theory, we may suppose that neither the aggressive centers nor the flight centers were able to discharge their action-specific energy sufficiently and the energy from these centers "spilled over" or "sparked over" to the feeding center. When a courting bird (for example, the yellow-eyed penguin) executes a vigorous display but the intended mate fails to respond appropriately, the animal will often start preening itself. In this case, we may suppose that mating energy "sparked over" to the preening center. Such irrelevant behaviors are called displacement behaviors. They are behaviors activated not by their own, or autochthonous, centers but by energy derived from other, or allochthonous, centers. As we shall see subsequently, this concept of displacement activities plays a considerable role in the ethological interpretation of social behavior. Here we need only note that it is a type of activity suggesting that action-specific energy can be accumulated in instinct centers. It should also be noted that Tinbergen and other ethologists now favor an interpretation of the origin of some displacement behaviors somewhat different from that one given above and based on the reduction of inhibition.

Related to displacement activity and similarly suggesting the presence of action-specific energy is the common tendency, described in Chapter 1, for a subordinate animal to pass along punishment it has received from its superior to a subordinate in the social hierarchy. Since such action is behavior-specific (autochthonous), it cannot be said to be a displacement activity in the technical sense, although of course the action is "displaced" to a new object. This passing along of an instinctive act to an object other than the one presenting the optimal stimulus has been called redirection. It occurs commonly with respect to sexual activity, parental care, and even feeding reactions. It, too, suggests the accumulation of action-specific energy with the lowering of thresholds for release of motor centers.

Evidence from many other sources may be adduced to support the concept of action-specific energy. For example, the development of a type of instinctive behavior in the life of an individual or its arousal in the course of the yearly cycle of behavior often shows evidence of the gradual development of a particular "mood" for such behavior (see Chapter 2 for examples of the development of such behaviors). The incipient movements characteristic of early discharge phases of a response are called "intention" movements. Thus we spoke earlier of territory establishment in song sparrows being shown briefly and in a desultory manner early in the season. Gradually such behavior is displayed with more and more vigor and in a more and more sustained manner. The intention movements pass over into the fully expressed instinct. Similarly, within the territory, the settling down to a final nest site is seen in many birds to follow a similar period of increasing intensity of exploration of nest sites and handling of nesting material. Intention movements are, of course, also shown in adults. Before flight, many large birds show flying-off intention movements. Such intention movements enter into social signaling in an interesting way which will be discussed later.

It is also noteworthy that physiological investigation of the nervous system gives indication of neurological centers from which some complex activities are controlled as units. As has been explained, certain areas of the hypothalamus have been found to act as stimulative or inhibitive centers of organized rage, sleep, or feeding behaviors (Fig. 7.1). Presumably these are principal parts of the apparatus reg-

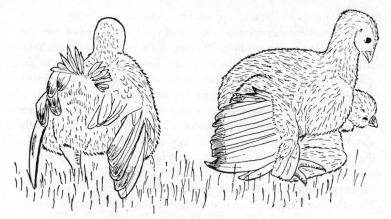

FIG. 7.1. Courting Behavior in Androgen-Injected Young Turkeys
Immature turkeys injected with male hormone carry through courting behavior in a highly organized fashion without previous experience. Note
that the animal on the left is strutting with wings lowered and tail fan (as
yet undeveloped) spread and turned toward the stimulus object (a mount
of a squatting turkey). The bird on the right is correctly oriented and is
copulating. The pattern of behavior is thus seen to be already organized
and ready for action in the immature bird. (After M. W. Schein and E. B.
Hale, *Animal Behaviour*, 7 [1959].)

ulating instincts which include these behaviors. Other neurological
studies have shown that certain regions of the central nervous system
maintain rhythmic discharges even when isolated. Respiratory and
locomotor regions of fish and amphibian nervous systems show such
self-maintained rhythmic activity in spite of apparent isolation from
sensory impulses. This physiological evidence of autonomous action
in neurological centers applies to the expression of relatively simple
activities that, at most, contribute to only the motor aspect of some
instincts. Perhaps the analogy to ethological concepts should not be
pushed too far. The centers conceived by ethologists involve, as we
shall see, at least additional mechanisms for the selection of stimuli
(Tinbergen, 1951).

We have referred to typical instincts as units of behavior because,
though composed of many individual acts, the behavior tends to hang
together as a unit. Many such behaviors, however, seem to fall into
two subdivisions; the first has been called appetitive behavior and refers to the generalized seeking or exploratory behavior that an animal
shows when an instinctual mood is first aroused in it. The stickleback,

as the reproductive season approaches, shows a wandering behavior accompanied by an exploration of nooks and crannies along the water's edge. We might say he is "looking" for a suitable territory. Similarly, the starling's search of the ceiling, described by Lorenz in the quotation above, constitutes appetitive behavior. Wallace Craig (1918), whose early studies first suggested this concept, wrote: "An appetite so far as externally observable is a state of agitation which continues so long as a certain stimulus is absent." Such behavior is strikingly variable in expression and emphatically gives the appearance of being "purposive," in that it continues indefinitely until what might appear as the "goal" has been realized. In fish, bird, or frog, searching continues restlessly until achievement of the goal brings it to an end. Thus the hawk explores and maneuvers until the chance to pounce on prey appears, and the frog turns to follow a crawling creature until the stimulus to flick his tongue in capture appears.

The consummatory act, on the other hand, is quite stereotyped and appropriately described as a fixed-action pattern. The flick of the frog's tongue is so simple there would not seem to be much occasion for variation. The same fixity applies, however, to rather complex actions, such as the food handling of the starling described above or the territory-defense behaviors in the fish or birds as we discussed it in Chapter 1. Some authors, Lorenz, for one, have suggested restricting the term instinctive behavior to these stereotyped consummatory acts. Indeed, if we are to insist that instinct cannot have any element of learning in it, this might seem appropriate; for, as we shall see later, learning enters conspicuously into appetitive behavior. However, the exclusion of the appetitive behavior from the instinct destroys the unity of the activity and makes it difficult to visualize the role of the activity in the life of the animal. From the point of view of the animal's survival, the appetitive behavior provides the seeking of necessary goals, and the consummatory behavior represents the achievement of the goal. An important reason for considering both as part of an instinct is that, in this concept, both derive their energy or drive from the same central mechanism. When that mechanism is depleted, both behaviors disappear.

From its nature it is obvious that appetitive behavior does not serve to discharge the action-specific energy characteristic of an activity for if it did the appetitive behavior would cease. It is the consummatory act which discharges action-specific energy. Thus, after the consummatory act is carried out, the appetitive behavior ceases, at least for a while, until the energy store accumulates again. After discharge, the animal appears "satiated" with respect to that behavior. This leads to the concept that the "goal" of the appetitive behavior is the expression of the consummatory act and not, as we are tempted to think, the satisfaction of a physiological need such as hunger, rest, etc. In other words, what the hungry hawk described above is "seeking" is the opportunity to pounce upon prey, not the food as such. The activity of pouncing depletes the instinct center and thus quiets the restless seeking behavior. Of course, the food that results from the successful hunt under natural conditions stills the hunger of the animal. In the absence of hunger the action-specific energy is not so quickly re-formed in the center, and consequently the physiologically satisfied animal does not return to the appetitive behavior of hunting as quickly as the unsuccessful animal. The satisfaction of the physiological need does play a role in the behavior, albeit a secondary one.

Evidence for this concept of the consummatory act as goal can be seen even in natural conditions. For example, the young passerine bird in the nest shows gaping activity; that is, it opens its mouth wide and stretches its neck upward whenever a parent lands on the edge of the nest. Ordinarily the bird gets fed at this time, but if a nest parasite such as the cowbird or cookoo is present, the parasite may get all the food. The gaping response of the other birds in the nest nonetheless diminishes considerably in the same way as if they were fed. In dogs, too, it has been found that the drinking action of a thirsty dog ceases after it drinks a certain amount. This occurs in a normal dog and in one in which a pipe has been inserted into the esophagus so that all the water swallowed escapes from the animal, and its thirst cannot be said to have been satisfied physiologically. Of course, the action-specific energy of the system is quickly re-formed in the deprived animal, whereas the sated one may not again be ready to discharge the consummatory act so soon. That the "goal" of an animal's appetitive behavior is the consummatory act, rather than the gratification of a specific "need," is one of the basic tenets of the modern concept of instinct. It is fundamentally this conceptualization which frees us from the mysticism that has clung to the older ideas of instinct. The notion that the goal of instinctual behavior is a physiological need inevitably implies purposefulness and rationality, even consciousness on the part of the animal. Even where objective methods of experimentation are employed, the true nature of the goal may be missed. As Tinbergen (1951, p. 106) says, "Even psychologists who have watched hundreds of rats running a maze rarely realize that, strictly speaking, it is not the litter or the food the animal is striving toward, but the performance itself of the maternal activities or eating." Perhaps Tinbergen should have gone further and pointed out that the maze-running activity itself is part of a territorial behavior that is "satisfied" only by the activity of exploring the maze. The drive to explore its environment is a goal of behavior in many animals, irrespective of any "material" reward that such exploration produces.

THE SENSORY COMPONENT OF INSTINCT

The ethological concept of instinct suggests that the neural energy for instinctual action is bound up in a neural center and the path of discharge laid out in advance; the only thing lacking is a stimulus from the environment that determines the timing of the discharge. From this point of view, the concept of the discharge of instinctual acts by simple sign stimuli or signals does not seem strange or unexpected. Yet it is this concept, more perhaps than any other, that is a basic novelty of the system proposed by the ethologists.

According to this concept, the discharge of each consummatory act is blocked by a mechanism which can be released only by stimuli reaching it from the animal's environment. The blocking mechanism has been called the internal releasing mechanism (IRM). The idea of release is emphasized in this terminology because, from the point of view of behavior, it is the release rather than the blocking that requires explanation. The stimuli from the environment that release most IRM's that have been studied are found to be simple and specific (Fig. 7.2). Each stimulus fits its IRM as a key fits a lock. That is to say, it contains characteristics which make it distinctive, characteristics which do not ordinarily occur in the environment except in the appropriate releasing object. Thus the mating response of an animal should be given only to an appropriate mate; the feeding response should be given only to appropriate food. Rationally considered, appropriate food for a frog might be said to be any organic material not specifically harmful. Experiment shows, however, that the frog snaps at any moving object of small size but not at stationary food. Consideration of the normal life of a frog shows, furthermore, that practically the only objects of appropriate size that move in its environment are small live organisms. These are, of course, appropriate food. As a practical matter, therefore, it is advantageous for a frog to

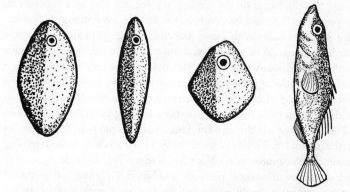

FIG. 7.2. Experimental models used in testing for the releaser of territorial defense behavior in the three-spined stickleback are shown here. The
right figure is an accurate representation of a male, but the belly is light
colored in contrast to that of the normal breeding male, which is red. The
other models depart to varying degrees from the normal appearance.
However, the ventral sides are colored red. Each of these elicited much
more attack from the territorial male than did the right figure. This experiment shows that it is the distinctive red belly rather than other characteristics of the rival which serves to release the fish's attack behavior. (After
N. Tinbergen, *Study of Instinct*, 1951.)

snap at any such object, for the combination of movement and small
size rarely occurs in inappropriate objects under normal circumstances. Of course, in the laboratory, we may wonder at the stupidity
of a frog starving "complacently" to death in spite of the fact that
pieces of fresh meat are put into its cage every day. The releaser for
the instinctual behavior here represents about as simple a combination
of characteristics as would practically "key-out" the normal nature
of its food in natural conditions.

It is the contention of this theory that the sign stimuli or signals
that release instinctual behaviors are combinations of characters, or
even single characteristics, that specifically identify the appropriate
object and do not occur ordinarily in inappropriate objects. The reaction is given to these characters and not to the situation as a whole.
To quote another example from the feeding mechanisms of animals
(for the moment we wish to avoid social responses), we may point
to the striking reaction of a rattlesnake in the dark to any small, warm
object that moves in its neighborhood. To a rattlesnake in its normal
habitat this can only be a small mammal, which of course is appropriate food. In the laboratory a warm electric bulb wrapped in cloth

elicits the discharge of the feeding mechanism just as effectively as proper food. The combination found to release the feeding mechanism of a parasitic tick—in this case, the drilling action of the proboscis— was a combination of warmth and butyric acid odor, a smell common to the skin of many mammals. The tick may remain quiescent on a bush for months or even years from one meal to another until a warm object smelling of butyric acid brushes by. The tick then releases its hold, drops on the object, and inserts its feeding month-parts (Lorenz, 1957). This again is the rare combination of simple stimuli that occurs in the normal tick's environment only in association with a passing mammal, the appropriate object for feeding for the tick. Specificity and simplicity are almost perfectly combined. Only the evil genius of the experimenter would smear butyric acid on a rock warmed in the sun to deceive a tick into ruining its proboscis.

There are a number of complications to the idea of the consummatory action which deserve some consideration. One is the fact that many actions which appear to be fairly simple have been shown to be composed of a series of separate actions. For example, the approaches of various insects to food or sex objects can be separated into at least two reactions in a chain. Bees and some moths are attracted to flowers by odor, color, and details of form, each stimulus being effective at a different distance (Baerends, 1950). Some male moths find the female by a response to odor from the female's scent glands which activates a reaction system. This induces the animal to fly upwind, a response which would, of course, normally bring it into the vicinity of the female giving off the scent. Some of the instances of sign stimuli may appear complex because a sequence of separate responses, each with its own appropriate releaser, may be mistaken for a single response.

Another elaboration of the concept of instinct, at one time emphasized by Tinbergen, is that instinctual behavior is organized into a hierarchy. The top levels of the hierarchy govern the more general activities. Energy released from them activates lower levels, governing more specific activities. For example, the top level of reproductive behavior in a stickleback fish sets up the appetitive behavior of inshore migration and searching for a territory. The releasers which act on this top center are presumed to be those which determine the localization of a territory. Though they are unknown in detail, they presumably include weeds and warmth. Once the territorial center has been activated by energy released from the general reproductive center, the animal shows territory-guarding behavior. This includes the appetitive behavior of searching through the area, especially patrolling near the boundaries. Should a male presenting a red belly swim into the territory, this would release fighting behavior; or contrariwise, should a female with a swollen white belly appear, courting behavior would be released. The release of the lower centers initiates the consummatory acts, which drain action-specific energy from their centers. Although this concept of a hierarchical organization is theoretically interesting, it does not seem to have been particularly helpful in experimental analysis and has been de-emphasized in recent ethological writings.

The Stimulus in Learned and Instinctual Reactions

The essential difference between the ethological concept of the stimuli of instinctual actions and those of learned reactions is the relative simplicity and fixity of the former. Lorenz (1950) accepts the concept that in learned reactions the stimulus is a *Gestalt,* or a whole. Such a stimulus is effective even if considerable detail is altered, as long as the over-all unity is maintained. For example, if an animal is trained to respond to a circle, its response is not disrupted if, instead of a continuous circle, a dotted outline is presented or if the axes of the figure are altered, within limits, to produce an ellipse. According to this notion, the perception upon which learning depends is the perception of a field, or whole, which may be given by a physically imperfect representation. In contrast to this, sign stimuli are characterized by the dependence upon single, or a few, aspects of the stimulus without regard to the loss of any resemblance to the wholeness of the normal stimulus. Thus the territorial defense behavior of the English robin is set off by the posing display of a rival. If this response were to a *gestalt* perception, we would expect that a dummy would serve to set off a response proportional to the over-all similarity of the dummy to a natural robin. Yet experiment reveals exactly the opposite. One can remove the head, the tail, the back feathers—in fact everything that makes it look like a robin (to the human observer)— and still elicit a good response as long as a bunch of red feathers is retained. On the other hand, if a stuffed, complete robin is presented but with the red breast dyed brown, no defense reactions are elicited from the territory owner. To our eye this model closely resembles a robin, but to the robin's releaser mechanism a simple bunch of red feathers on a wire is much more effective (Lack, 1943).

We need not attempt here to go into the theory of perception. It is obvious that some configurational characteristics of a limited kind may play a role in releaser perception, as indeed Tinbergen (1953) finds in the escape response of fowl to the outline of a flying hawk, for example. Here it is not only the shape of the figure that releases the escape response but the shape in relation to the direction of movement. Even if we accept the observations of Tinbergen in this regard (they have been questioned by other experimenters), it is obvious that we are dealing here with a simple perception which ignores much of the characteristics of the stimulus object. In any event the relative simplicity of releasers and the absence of strong *gestalt* tendencies in perception are outstanding characteristics of instinct releasers. We shall see that these aspects are of great significance for their role in social communication.

Different elements of the releaser have their separate effects, so that they summate and together are more effective than each aspect is by itself. This phenomenon of summation is discussed more extensively by Tinbergen from the point of view of the evolution of signaling systems. We may note further that the naturally occurring releasers may not be maximal in effectiveness with regard to each component characteristic. As a consequence, it is possible to provide artificial stimuli that are even more effective than the natural ones. Thus certain birds, when given the opportunity, will choose to incubate monstrously large or excessively spotted eggs in preference to their own. These so-called supernormal stimuli give striking evidence of the mechanical nature of the animal's responses to releasers.

Imprinting

We have not said anything hitherto with regard to the question of whether the stimulus recognition is innate or learned. It was tacitly assumed that the organism, somewhere between the sense organs and the IRM, has an innate capacity for recognizing appropriate sign stimuli. It would be difficult to see how such a sign stimulus as a red breast could be learned, since the young robin does not see the reaction performed before it is itself capable of carrying it out—in the first springtime of its life. An element of learning does enter into stimulus recognition of some IRM's. This was discovered in experiences with incubator-hatched geese and other fowl. It was found that if the young birds are handled by the experimenter during their first few hours of life, they will thereafter react to him and to other human beings as they normally would to their parents. In geese, for example, the young will faithfully follow the experimenter wherever he goes. They will seek the protection of his body whenever they hear the normal warning cry of the species. This phenomenon of the learning of a releaser stimulus in a short exposure to it has been called imprinting (Lorenz, 1957). It was discussed briefly in Chapter 2 as an aspect of parental care.

Imprinting has many characteristics which distinguish it from ordinary learning processes. In fact, Lorenz has often insisted that it cannot be considered to be learning. Since, however, imprinting is clearly a behavioral effect of experience, it seems arbitrary to distinguish it from other learning. We shall consider it a type of learning and note the differences from other kinds. One of these distinguishing characteristics is that the capacity for it is limited to a brief, specific "critical period" of an animal's life. If imprinting is not accomplished during the first few days of a hatchling's existence, it will not "take" at any other time. Such an animal will fail to respond appropriately to any other animal, and no amount of association with members of its species will bring out the response of following. Chicks have been found to imprint to one another as well as to the hen in the first seventy-two hours of their lives (Guiton, 1959). Hess (1962) finds that a principal factor in determining the strength of imprinting is the effort expended by the young in the process. He also emphasizes the differences from ordinary reward-conditioned learning (Fig. 7.3). Most remarkable, indeed, is the fact that when the animal which has been imprinted to a human being becomes mature many months later, it will show courting responses to humans even in preference to its own species. It is thus clear that imprinting can take place long before the reaction with which it is associated matures. This is a phenomenon often observed in birds. Hand-raised birds commonly display courting behaviors to the trainer's hands (Schein and Hale, 1959).

Not all the instinctive reactions of a given organism are directed to the same imprinted object. For example, Lorenz (1957) describes how jackdaws imprinted to himself would court him but would show

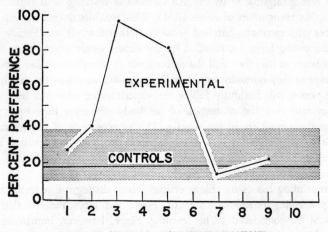

FIG. 7.3. When inexperienced chicks are presented with a choice of a white triangle on a green background or a white circle on a blue background, they show a small preference for pecking at circles. The triangles were selected only about 19 per cent of the time in tests of about four hundred chicks. Groups of chicks were exposed to these stimuli at various ages from one to ten days, and their pecks at the triangles were reinforced by food rewards. When tested later against these same stimuli the effect of their early experience was shown only by animals receiving reinforced experience on the third and fifth days. This demonstrates the existence of a critical period at three to five days for learning the pecking response. (After E. Hess, E. Bliss [ed.], *Roots of Behavior*, 1962.)

flying-off behavior only to crows and would attack him as a predator when he held a black waving object, the sign stimulus equivalent to a struggling jackdaw, in his hand. Each reaction had its own releaser. Under natural conditions the innate releasers (for predator attack) and those dependent upon imprinting (for courting) would both be given to jackdaws only, since an animal raised by its parents would be so imprinted.

Another characteristic of the imprinted releaser is that it is of a general, supra-individual character. That is, imprinting is to the human or other species, not to one individual human being. Again this is most strikingly shown in courting behavior. Instances have been recorded of male turkeys imprinted to a male caretaker, not only showing courting behavior to him, but showing rivalry and antagonistic behavior to women, presumably because the loose clothing of the women suggested male turkey characteristics, such as the wattles. This long-term and very stable characteristic of imprinted stimuli is of utmost importance, because, as we shall see, it affords a possible explanation of many aspects of socialization in animals. It should be noted, however, that the generalized bond established by imprinting may be refined and restricted to one individual by later learning processes. Thus, though ducklings at first follow any duck, they normally learn soon to restrict their following response to their own mothers.

Specificity of Learning Processes

It is not our intention here to attempt to consider the nature of animal learning generally but simply to confine our discussion to certain specific aspects that are more or less directly applicable to the problem of socialization of animals. One of these is that animal learning often shows a high degree of specialization. It thus departs from the common conception of learning ability as a generalized phenomenon, a single capacity which may be called intelligence. We usually assume, further, that there is a clear-cut correlation between the degree of development of the brain and the various aspects of intelligence, such as memory, rate of learning, complexity of learning, and insight learning. In a broad phylogenetic view, this concept may indeed be justified, for it would certainly appear that annelid worms are inferior to fish in any reasonable test and, similarly, that mammals are correspondingly superior to amphibians. The nervous system has made such enormous strides between these grades of animal organization that the variations to be found within each group and in response to different tests may be ignored in a broad comparison between the stages. However, when we ask specific questions concerning learning capacity, we are immediately led to realize that there is another factor, namely, ecological adaptation, which plays a very important role and which may override the broad phylogenetic progression in mental capacity.

If, for example, we ask about memory, we not only find that some birds such as parrots, are credited (on reliable evidence) with memory capacities for individual recognition extending over years and comparable to the highest ability shown by mammals, but we find evidence of something comparable in fish. For example, salmon are known to be able to recognize in some way the stream in which they were hatched. They return to it for spawning after five years or more of life in the open ocean. Is this ordinary (conditioned) learning or imprinting? Does it depend upon a memory trace (engram) in the nervous system? We do not know, but it would appear highly arbitrary to deny that it is memory of a kind. If we consider, as another example, the learning of localities and pathways, as in maze learning, we find great discrepancies between phylogenetic predictions that capacities increase with size of brain and actual performance. Wasps, whose small nervous systems must be morphologically very simple compared to that of vertebrates, nevertheless show an unexceeded ability to learn a homing response. Various digger wasps, for example, after excavating a burrow, cover it carefully, completely concealing the opening. They then circle over it once or twice and are off on a hunting expedition to stock their nest holes. They may not return for hours or even days and yet have no difficulty in returning to the exact spot. Experiments show that this capacity depends upon the learning of visual cues furnished by neighboring landmarks. Such learning appears to be achieved in a brief orientation flight and to persist for days. The human observer often confesses his inability to equal this achievement (Thorpe, 1956). Since song birds tend to return to the neighborhood of their birth and to the very territory held the previous year, it is obvious that they too retain a memory for the necessary details, not only of the topography of the territory itself, but of the surrounding country as well. It is clear, of course, that these "mental" capacities are part of the adaptation of the animal to its particular mode of life and do not imply a general long memory or capacity for learning complex patterns other than those involved in the particular functional activity. Indeed, it may be that many other of the learning processes commonly studied in the laboratory by experimentalists also represent similar isolates and therefore give entirely erroneous impressions of the mental capacities of the subjects. For example, rats have been used very extensively in maze-learning tests. It is known that the rat's ability for maze learning is about the same as a human's ability to learn a pencil maze when blindfolded. Should this be taken as a general measure of learning capacity, even of capacity of one general kind? Rats under natural conditions have a tremendous exploratory drive, which leads them to explore every new aspect of their territory. Under such conditions rats come to know their territories very thoroughly. The biological usefulness of this knowledge becomes apparent when a predator, such as an owl, appears. The rodent then shows a capacity for dashing for the nearest hiding place without the least hesitancy or circumambulation, irrespective of its position at the moment. Knowledge of the topography of their home range is thus of overwhelming importance for the survival of rats. The high capacity for maze learning is seen, therefore, to represent a specialized ability of the rat and can hardly be expected to bear any close relation to the general capacities of its level of nervous organization. Incidentally, it may be noted that what psychologists call "latent" learning of mazes —that is, learning which takes place when rats are simply left in a maze and permitted to explore without being given a reward—appears, from the ecologists viewpoint, to be learning that is highly rewarded. It is obvious that knowledge of its environment is one of the primary "needs" of a rat, and the achievement of that knowledge by exploration thus satisfies just as food satisfies hunger.

The biologist likewise tends to take a skeptical view of the phylogenetic interpretation of many other kinds of comparative learning capacities which are not considered in relation to the animal's mode of living. For example, psychological experimentation indicates that lower animals, even apes, have very limited capacities for carrying through delayed responses, a few minutes being top score in the usual laboratory test. In the laboratory situation the facts are clear enough. Yet, when one considers that many animals, birds as well as mammals, carry through activities involving long delays such as returning to the site of a kill or buried or hidden food, it is difficult to believe that short delays really represent the limits of the mental capacities of animals (Thorpe, 1956). One animal, such as the macaque monkey, may be particularly good in a type of problem such as the "Umweg" problem, that is, a problem requiring that the subject go in a roundabout way to the solution. This capacity may indicate only that that animal normally lives in an environment—amid trees, for instance—where the roundabout path is frequently the only solution to naturally occurring problems of obtaining food. Therefore, monkeys do well in such problems, and dogs do not. It may be that this difference is related more to their ecology than to the evolutionary status of their nervous systems.

This characteristic of animals, whereby their abilities for the learning of particular types of responses are related to their ecological requirements, is one of the fundamental points of emphasis of ethological studies. Tinbergen (1951) refers to it as a function of differences in the "innate disposition to learn." This specialized character of animal learning illustrates the fact that learning ability often depends on special innate factors, just as the expression of instincts frequently depends upon special factors of experience. Thus, an animal's behavioral characteristics do not segregate themselves strictly into innate or learned categories. Related animals with similar brain structures may differ considerably in their innate disposition to learn particular items (Fig. 7.4).

Our conclusion from the foregoing discussion is that mental capacities of learning and related phenomena correlate in details better with ecology then with phylogenetic status. From the point of view of social behavior, this concept is important. It indicates to us that specific items of learning, such as individual recognition, locality and homing learning, memory for early experiences, etc., can be surprisingly complex even in animals which appear to be otherwise very limited in their mental capacities. Thus we cannot, on phylogenetic grounds, consider that a phenomenon cannot be learned in a particular animal because it is too complex for the animal's grade of organization. The common phylogenetic generalizations about learning, such as that insects and birds are largely animals of instinct, whereas mammals show much more learning in their behavior, cannot be applied a priori to any particular activity with confidence. Indeed, Hinde (1961) has questioned the verity of the common opinion of the superior learning capacity of the mammalian, as compared to the bird brain.

Motor Patterns in Learning

Much learning leaves motor patterns unmodified but simply associates such patterns with new environmental stimuli. It is, for example, obvious from what was said above about maze learning in rats that maze exploration is a motor pattern that does not need to be taught to rats but comes to them "naturally," presumably innately. Certain other animals, such as chickens, have been used in maze-learning experiments but without much success, because maze running is not a normal part of their mode of life. Sometimes the experimenter went so far as to use electric shock or blasts of air to move the animal along in the maze. Hediger (1955), in his discussions of the training of circus animals, emphasizes the use of natural capacities. Sea lions, for example, hardly need any teaching to balance balls on their snouts; they will of their own accord play with floating pieces of wood by tossing and catching. Apparently, flexibility of neck correlated with

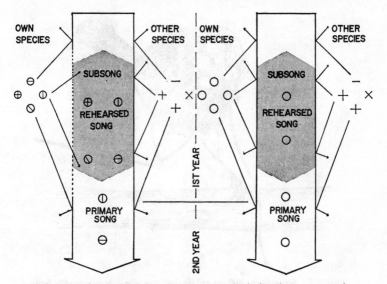

FIG. 7.4. The singing of various passerine birds has been extensively studied in birds raised in isolation and variously exposed to the song of their own or other species. During the first year the birds show some singing and call notes but not the distinctive song of the species. This is called subsong. It is followed, usually by the fall of the first year, with the introduction of some motifs of the characteristic song. The following spring, the definitive song of the species, the primary song, is produced, usually after a brief period during which subsong and rehearsed song are recapitulated. In some species, as the corn bunting, the evidence indicates that this developmental process proceeds effectively, independently of any learning from other birds. This is indicated in the *right diagram* where the arrows from the symbols for motifs of other species or of its own species are shown as reflected without effect on the developmental process. This condition is regarded as probably the primitive one in song birds. Most present day birds such as the blackbird, white throat, canary, and chaffinch, however, show a period of susceptibility to the influence of other members of their own species. In this case, the motifs of other species are shown as rejected, but those of their own species as penetrating and having an effect during the susceptible period (*left diagram*). In those birds refinements of the motifs are learned from other members of the species, and the young of these species complete their full repertoire of primary song by learning. Some species as the red-backed shrike are able to add motifs from other species and thus are true mimics. (Modified after Lanyon, *in* W. E. Lanyon and W. Tavolga (eds.), *Animal Sounds and Communication*, 1960.)

FIG. 7.5. Some birds, such as goldfinches and jays (above), readily learn to secure food attached to strings by pulling up the strings with their feet. Others, such as robins and wrens, do not manipulate the strings and do not learn the trick. To an important extent the ability to solve this problem depends upon innate motor patterns for manipulating objects with the feet. In some cases these are so well developed that very little experience seems necessary and the learning achievement takes on the character of "insight." In others previous experience is an important factor in the animal's ability in learning this trick. (After W. H. Thorpe, *Learning and Instinct in Animals*, 1956.)

precise head movements controlled by binocular vision is part of their equipment for hunting fish. They tend to use these capacities under other conditions. Training of animals proceeds most successfully when it confines itself to using the motor equipment used by the animal in its normal mode of living. When we speak of a reaction as learned, we may be referring only to one aspect of it—the association between stimulus and response—but the response itself may have an innately determined co-ordination of parts (Fig. 7.5). This is not to deny that genuinely new patterns of motor co-ordination may also be learned. We wish simply to emphasize the other side of the coin discussed before. Just as previously we stressed the idea that the fact that a response appears to be innate does not preclude the existence of elements in it that depend upon experience, so we wish now to point up the fact that because a response is commonly considered learned, this does not mean that large elements of it are not determined or restricted by innate factors.

Social Releasers and Kumpan Theory

The ethological theory of instinct has given us a new insight into the nature of animal socialization and society. The basic concept here is that animals carry in their own structure and behavior the releasers that evoke appropriate social responses in their fellow group members. Such releasers are called social releasers. As an example of this, we may mention the fact discussed in Chapter 1—that, in territorial animals, males arouse antagonism of other territory-holding males whenever they enter their domain. Experimental analysis of the way in which the territory holder recognizes his rival has shown that this is dependent in many birds and fishes upon relatively simple signals. For example, in the English robin, the red breast is the primary sign stimulus for the attack (Fig. 7.6). The animal thus bears on its own body the signal to release the appropriate social response of its rival.

Simple morphological sign stimuli are usually supplemented by peculiarities of behavior. Some of these, such as the posturing and singing from high perches by song birds, may be considered as ways of displaying the animal's distinctive morphological and auditory characteristics. Others, such as bill-clappering of cranes or the deep bow followed by hissing and snapping of herons, may not be dependent upon special physical characters for their effectiveness. At any rate, all of such structures and behaviors serve to elicit appropriate behaviors in another member of the species. If this be a male, they may serve to intimidate him to leave or, contrariwise, arouse him to fight for the territory. The female's response to the same signals may be quite different, at least, if she has reached an appropriately advanced stage in sexual development. She may respond by passive but insistent staying in the territory, in spite of the attacks of the male. Here we see another characteristic of social releasers. They act in situation-specific situations and are responded to according to the respondent's condition or mood. Thus the same sign stimuli may arouse territory fight, flight or courtship behavior, or, from a young bird, no particular response at all. Two animals, which may be distinguished as the "sender" and the "receiver," in analogy to radio transmission of signals, stand in one particular relation to each other

FIG. 7.6. Threat display of a robin redbreast (*right*) at a stuffed specimen (*left*) mounted in his territory. The red breast which is the releasing signal is made conspicuous by this posture. (After D. Lack, *The Life of the Robin*, 1943.)

with regard to a particular sign stimulus. This relation may be the mate-relation or rival-relation. They may be first one, then another; a female first unready for mating may respond as a rival and later as a mate. As human beings may be companions to one another in bowling or in business without being related in other ways, so animal companions, when their relation depends upon sign stimuli, bear only situation-specific relations to one another. The German students of behavior use the German term "kumpan" for such a situation-specific companion. The word in German apparently more nearly carries that connotation than does the English equivalent. It would, therefore, be convenient to retain the term for a situation-specific companion. It is at once technically accurate and satisfyingly familiar for our purpose.[1] The fundamental role of releasers in social behavior then may be summarized as follows: Animal kumpans furnish to one another the appropriate sign stimuli to release the behaviors adequate for a particular social interaction between them. These interactions are

---

[1] Anglicized plural—kumpans.

specific: male for female, rival for rival, parent for child, flight companion for flight companion, etc. Unlike learned reactions which involve the recognition of the individuality of each animal as in the dominance hierarchy, the kumpan functions merely as the bearer of the releaser for a particular behavior and is not recognized as an individual (Lorenz, 1950, 1957; Tinbergen, 1951, 1953).

Evolutionary Trends in Social Releasers

One of the most significant aspects of social releasers and one in which they differ from non-social releasers is that, being parts of organisms, they are subject to evolutionary change (Fig. 7.7). The releaser mechanisms are subject to selection as are other characteristics of the organisms. There is, therefore, a tendency for them to evolve into more and more effective types. This evolution proceeds in both members *pari passu,* for the "sender" and the "receiver" must always be in tune with each other. This lock-step evolutionary process permits near perfection to be attained and accounts for many of the extraordinary specializations and varieties of pattern, color, and sound that animals have evolved as signals.

Effectiveness implies a number of characteristics. One of these is distinctiveness. That is, the releaser will be effective in so far as it is readily and clearly differentiated from those of other species or other reactions in the same species. If it is distinctive, it will not evoke its action in an inappropriate kumpan. For this reason, we find that movements and structures serving as releasers tend to diverge from one another in evolution much as ordinary structures do. From a common, original type they tend to become separated into a variety of specialized forms that distinguish each species from the others. Thus each species of song bird has a distinctive song or other display character that prevents confusion between species.

Besides distinctiveness, another evolutionary trend to be found in social releasers is that of developing conspicuousness (Fig. 7.8). We have seen many examples of this with respect to courtship behaviors, such as those of the peacock; although in these cases we inferred that physiological co-ordination of mates was also involved (see Chapter 2). Simpler examples of releaser function are found in the many characteristics by which group-living animals signal danger as they flee. When one bird flies up, the rest of the flock usually follows. Many

FIG. 7.7. Some signaling movements originated in more elementary behavior patterns such as attacking, escaping, mating, and nest-building. Upright threat (*top*) consists of the assumption of an upright posture at the initiation of the attack, head held high for downward peck and wings partly opened for striking. If the opponent does not yield to the threat posture, the bird may attack, or, if the attack is inhibited by fear, the bird may turn sidewise in an appeasement posture, turning the beak upward.

Grass-pulling (*middle*) is often seen in territorial disputes. The bird pecks furiously at the grass as though attacking an opponent. This is a "redirected" attack. The movement terminates with a sideways flick of the head, such as is seen in normal nest-building. This element of the behavior is thus a displaced nest-building movement.

The kittiwake, a small gull, advertises the nest site to passing females by "choking," which consists of bending over followed by rhythmic up-and-down movements of the head (*bottom*). These are movements normally performed during incubation and infant feeding. They have here become ritualized as advertisements of territorial status. (After N. Tinbergen, *Scientific American*, July, 1962.)

ducks and geese, in which this characteristic is strongly developed and important in keeping the flock together, show conspicuously developed stripes or other markings on their wings. These become visible only when the animals take off. Such "specula," as they are called, constitute flashing signals that appear to release the flying-up behavior of their flight kumpans. A similar phenomenon in mammals is

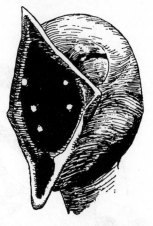

FIG. 7.8. The nestling of the bearded tit illustrates the remarkable development of conspicuous signals on the inside of the mouth which serve as releasers of parental feeding responses. There are four rows of pearly white conical projections set against a background of black. The interior is framed by a carnelian red border and lemon-yellow gape-wattles at the corners of the mouth. (Drawn from a photograph.)

the white rump and under-tail hair of ungulates, such as many deer and antelope. Our western pronghorn antelope shows this development to an extreme. The entire rump is covered with long white hairs. When the animal flees, these are stiffly erect, producing a bright patch that flashes in the sun as the animal dashes off (Fig. 1.1). This signal is said to be visible for long distances and to give warning to other members of the herd of the presence of a source of danger (Seton, 1953).

Of great interest for the general concept of social behavior is the way in which social releasers tend to evolve conspicuousness in behavioral movements by a process that has come to be called ritualization (Baerends, 1950; Hinde and Tinbergen, 1958; Blest, 1961). In this process the behavioral movements often take on a formal stiffness, slowness, or exaggeration that serves no function except to make them stand out sharply from ordinary functional movements. We have described previously the formal bowing, erection of head and neck, and bill-fencing that characterizes the courting and nest-relief ceremonies of gannets and other birds (Fig. 7.9). The "proud" postures with head held high and stiff walk of so many dominant animals, such as the master buck in the Indian antelope, may be viewed as ritualized threats (Fig. 7.10). Many of the intention movements mentioned earlier tend to evolve by ritualization (Daanje, 1950). Some threat postures, because of their exaggeration and slowness compared to normal fight reactions, serve as forms of communication. The stance, the carriage of the tail, and the facial expression of the wolf have been found to serve as signals of the status of the animal with regard

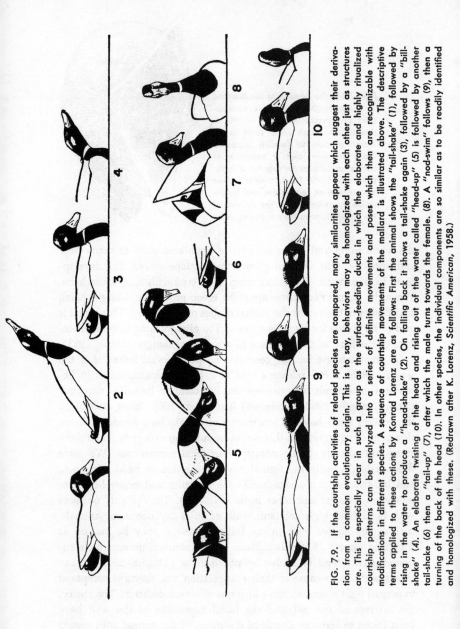

FIG. 7.9. If the courtship activities of related species are compared, many similarities appear which suggest their derivation from a common evolutionary origin. This is to say, behaviors may be homologized with each other just as structures are. This is especially clear in such a group as the surface-feeding ducks in which the elaborate and highly ritualized courtship patterns can be analyzed into a series of definite movements and poses which then are recognizable with modifications in different species. A sequence of courtship movements of the mallard is illustrated above. The descriptive terms applied to these actions by Konrad Lorenz are as follows: First the animal shows the "tail-shake" (1), followed by rising in the water to produce a "head-shake" (2). On falling back it shows a tail-shake again (3), followed by a "bill-shake" (4). An elaborate twisting of the head and rising out of the water called "head-up" (5) is followed by another tail-shake (6) then a "tail-up" (7), after which the male turns towards the female. (8). A "nod-swim" follows (9), then a turning of the back of the head (10). In other species, the individual components are so similar as to be readily identified and homologized with these. (Redrawn after K. Lorenz, Scientific American, 1958.)

FIG. 7.10. The master buck in the herd of Indian antelope leads the estrus female away from the others by pushing with the underside of his
neck. When they get to the mating arena he courts her by a neck-stretching
movement which appears to be a ritualized form of the herding movement
by which he led her there. He does not actually touch her in this neck-
stretching but repeatedly poses over her. The estrus female responds to
the male's overtures by standing quietly, whereas the non-estrus animal
moves away whenever approached by the master buck. Only after much
courting does the male attempt mounting. (From W. Etkin, "Dominance
Behavior in Black Buck" [motion picture].)

to dominance. These may be regarded as more or less ritualized autochthonous movements of dominance or submission. To some extent
they are ritualized intention movements of attack or flight. (See Fig.
10.11 in Chapter 4.)

In discussing the concept of action-specific energy, we mentioned
the idea that when an instinctual center accumulates large amounts
of such energy, it tends to be released either by vacuum activity or by
"sparking over" to some other, usually related, motor center. Such
sparking over we called "displacement." One of the most interesting
aspects of the contemporary study of instinct is the discovery of the
ways in which displacement activities tend to evolve into conspicuous
sign stimuli by ritualization. One of the best examples of this is the
analysis of displacement digging in sticklebacks. Tinbergen found that
when patrolling his territory borders, a male often turns vertically
and makes displacement movements of pecking at the floor. This was

seen when he was faced by a formidable opponent, usually the neighboring territory holder. When the experimenter exaggerated the conditions favorable for this display by crowding the males, the displacement activity became vigorous and frequent enough to reveal its true nature. The animals were not merely picking up bits of the substrate, as though feeding or randomly discharging, but they actually excavated nests just as in true nest building. Tinbergen's interpretation of this is that when the animal is faced with an opponent, its fighting behavior is released; but when the opponent is outside the territory or favored otherwise, the discharge of aggressive fighting is inhibited and tends to spark over to the nest-digging activity, which is another prominent element of the animal's territorial behavior. In the stickleback, this displaced digging activity has evolved into a sign stimulus which serves to inhibit the aggressive activity of the kumpan, with the result that the mutual display of displacement digging keeps the aggressive tendencies of the territorial males from becoming self-destructive by excess fighting (Tinbergen, 1951).

From our present point of view the significant finding is that the displacement activity has become a signal between kumpans. As such, its efficiency has been sharpened by natural selective action leading to ritualization. Such evolution has been in the direction of making the movements stand out by their conspicuousness. The threatening stickleback turns so that the opponent sees his side view; he erects his ventral spines, but, most conspicuously of all, he pecks furiously at the sand in contrast to the mild functional level of pecking in real nest building. This vigor is evident in many other examples of displacement activity. Herring gulls and greylag geese show grass-pulling displacement activity when fighting. This activity is expressed with much more "animus" than when similar action is taken in nest-building activity. The food-pecking activity that roosters often show as a displacement activity when fighting is similar.

Not only are such displacement activities made conspicuous by the extra vigor and by special postures, but it should be noted that the very inappropriateness of the action commands attention. To the examples mentioned above we can add the displacement activity of sleeping shown by many wading birds, such as avocets, in conflict situations.

In Tinbergen's view, many of the individual movements that constitute the courtship sequence of the stickleback are displacement activities. They are brought on by a conflict of drives in the male but serve as signals for the release of appropriate behaviors by the female at each point. Tinbergen regards the zigzag dance by which the male leads the female to the nest as an alternation of attack movements toward the female, who is violating his territory, and retreat movements toward the nest, determined by sexual releasing stimuli from the female. When he arrives at the nest, his posing with head partly in nest and later his poking of the female when she is partly in the nest are interpreted as displacement of the normal functional fanning movements by which the male aerates the eggs in the nest. Tinbergen regards these courtship movements as being derived from related activities of territory guarding and parental care which have been displaced and ritualized to form signals that release the appropriate behavior of the kumpan. The attractiveness of this hypothesis in providing a clear evolutionary explanation of these otherwise strange courtship behaviors is evident.

It should be noted that derived movements, whether developed from intention movements or displaced activities, tend to take on the same regular and fixed character noted in instinctual activities in general. Thus each species tends to have fixed forms for the displaced activity (avocets sleep, geese pull grass, etc.) when fighting is inhibited. In other words, the process of sparking over is not random, happening now in one way, now in another. In part, we can see that this is the result of the fact that the outflow of energy in displacement is to another activity related to the first—as nest building is related to territory defense in the stickleback. Of course, such fixity of response facilitates the evolution of the response as a social releaser; for, to function efficiently as a stimulus in social communication, there must be a one-to-one correspondence between the displaced activity and the activity it is displacing.

This summary of the evolution of animal signals is based on the earlier theoretic writings of the ethologists. In reading the next chapter, it will be noticed that Tinbergen now gives greater prominence to intention and other derived movements than to displacement. Though recognizing the importance of displacement, Armstrong (1950), Hinde (1961), and Rowell (1961), as well as Tinbergen, seek other explanations of the physiology of such behavior than that of overflow of action-specific energy as described above.

Learning and Socialization

In the discussion of kumpan theory above we arrived at the concept that animals are in some cases held together in appropriate social interactions by releasers, either innate or imprinted, which evoke appropriate interactions between them. Such a concept can apply, of course, only to the highly stereotyped behaviors characteristic of instincts. Examples of such behavior are readily found in lower vertebrates and were extensively commented on above.

There are many other behaviors to which the kumpan idea cannot be applied. These include the less stereotyped social behaviors, particularly common in mammals but found also in lower vertebrates wherever individuality and specificity in social relations are common. For example, in Chapter 1 we pointed out that social organization based on social dominance with its correlates of hierarchy and closed-group formation depends upon recognition of individuals. Pair formation likewise is based upon the ability of the mates to respond differentially to each other as individuals. Similarly, territorial behavior requires recognition of specific areas. These are all characteristics which must be learned in the course of an individual's own lifetime. Early experiences in a group are often of great importance in socialization in animals (Figs. 7.11, 7.12, 7.13, and 7.14). We may characterize such learning as familiarization with the environment, social and physical. Its importance in social behavior was sufficiently emphasized earlier in this book. Here we may ask how familiarization is attained and maintained.

We may note first that the capacity for each type of familiarization is genetically determined as a species characteristic. Thus each species has, as we pointed out above, its own innate propensities to learn in this respect. In birds with well-developed social dominance, individual recognition is quickly learned and long maintained, and similar relationships were seen to obtain in territoriality, mate recognition, and parent-offspring relations, when these phenomena were discussed earlier. Thus an essential factor in this type of socialization is the sensitivity of the animal to the particular learning experience.

Related to this sensitivity is another characteristic of the behavior of animals in which inter-individual familiarization is important. Animals commonly display behaviors which serve to reinforce learned relations. We have already emphasized this aspect in dominance behavior, by pointing to the patterns of dominance display behavior shown by the alpha animal in hierarchies (Chap. 1). We also interpreted mutual greeting or post-nuptial courtship ceremonies as tend-

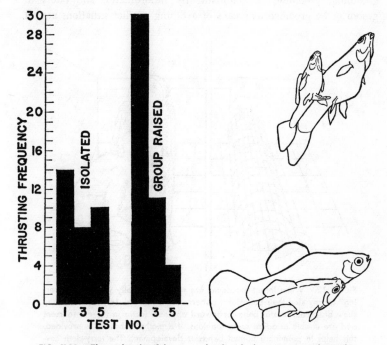

FIG. 7.11. The male platyfish courts the female by swimming alongside her and thrusting his gonopodium toward her genital aperture with the aid of the pelvic fin (*upper right*). Copulation is attained only occasionally in contacts which last several seconds (*lower right*). Males raised in isolation from species mates showed the essentials of courtship behavior but were partially inhibited in the expressions of some of their courtship activities in their first tests but not later. (After Shaw, *in* E. Bliss (ed.), *Roots of Behavior*, 1962.)

ing to maintain the learned relationship of mates (Chapter 2). In addition to these there are other forms of behavior that appear to function importantly in familiarization and which we have not yet discussed.

Some species of monkeys and other primates keep themselves quite free of parasites, debris, etc., by picking assiduously through their own and their companions' fur with their fingers and removing foreign matter. This grooming activity, especially between members of the group, seems far more extensively pursued than necessary for its toilet function. The animals are quite content to groom one another for considerable periods without receiving any tangible reward. Mutual grooming, grooming of dominants by subordinates, and offers to groom or be groomed as means of avoiding conflict situations are all

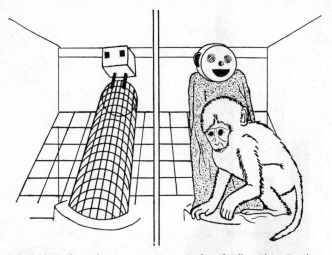

FIG. 7.12. When infant macaques are raised artificially without "mothering" of any kind their behavioral patterns are abnormal. In particular, they show disturbed, fearful behavior toward any new factor in the environment and are unable to adjust to companions. If a mother substitute is provided, this helps in permitting normal behavior development. The terry-cloth covered mother-surrogate (right) affords some comfort in fearful situations, but the wire mesh model (left) apparently does not. This is true whether the infant receives its milk bottle at the wire or cloth model. If raised with another infant, the animals cling together and thereby achieve considerable normalization of their behavior development. (After H. Harlow, in E. Bliss (ed.), Roots of Behavior, 1962.

commonplace among monkeys and apes. Some birds (for example, the yellow-eyed penguins) show a similar if less extensive mutual preening ceremony; but, with a few such exceptions, grooming is a distinctive behavior of primates (Nissen, 1951).

Such grooming resembles other types of persistent exchange of attentions that characterize many learned relations of familiarity in animals. The billing and cooing of doves and other birds, the persistent sniffing of one another by dogs and rodents, and the repeated

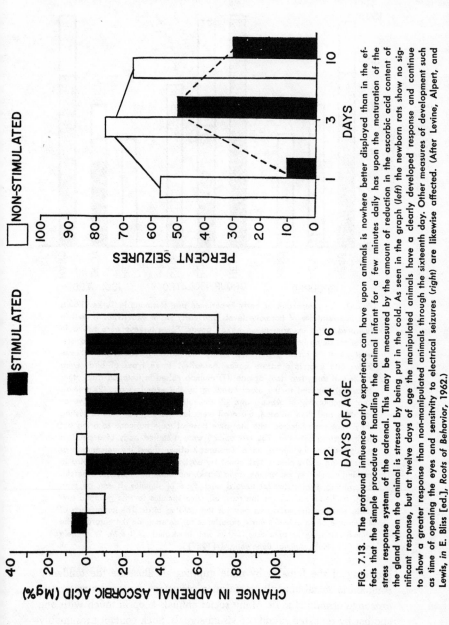

FIG. 7.13. The profound influence early experience can have upon animals is nowhere better displayed than in the effects that the simple procedure of handling the animal infant for a few minutes daily has upon the maturation of the stress response system of the adrenal. This may be measured by the amount of reduction in the ascorbic acid content of the gland when the animal is stressed by being put in the cold. As seen in the graph (left) the newborn rats show no significant response, but at twelve days of age the manipulated animals have a clearly developed response and continue to show a greater response than non-manipulated animals through the sixteenth day. Other measures of development such as time of opening the eyes and sensitivity to electrical seizures (right) are likewise affected. (After Levine, Alpert, and Lewis, in E. Bliss [ed.], Roots of Behavior, 1962.)

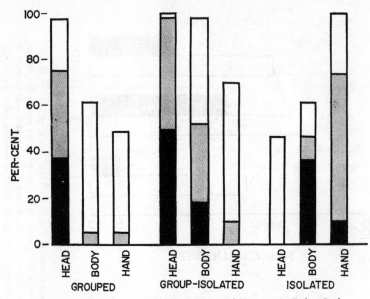

FIG. 7.14. Interaction of Early Experience and Hormone in Turkey Poults
The interrelations of hormone levels and early social experience is well illustrated in an experiment on turkey poults. Some turkeys were raised in
isolation, being hatched from eggs covered by aluminum containers to prevent visual or tactile experience during the hatching process in the incu-
bator and kept in separate boxes thereafter. Three types of birds were
studied at thirty-two days of age: (1) animals raised in isolation from other
birds but exposed to the usual handling by the experimenter, (2) animals
kept continuously in groups, and (3) animals raised in groups for five or
ten days and then isolated. Since all were immature at the time of testing
they were each injected with the same dose of male hormone to bring out
their courting behavior. The test objects were a stuffed body of a poult, a
poult head, or the human hand. The graph above shows the percentage of
time during a five-minute test, spent by each type of experimental bird in
making copulatory movements (*solid Black*), mounting attempts (*cross hatch*),
or strutting display (*white*) toward each type of stimulus. It can be seen
that the head was by far the most effective stimulus for the grouped animals, whereas the hand was best for the isolated birds. The importance of
the nature of an animal's early experience for determining the nature of the
stimuli releasing its behavior, is thus well illustrated. (After M. W. Schein
and E. B. Hale, *Animal Behaviour*, 7 [1959].)

prodding of the females by male guppies all illustrate the tendency
common in social animals to maintain persistent contact by reiterated
exchange of mild stimuli. Many social animals keep in touch with one
another by repeated sounds or visual signals. Such contact presumably

helps to maintain the sense of familiarity so important to their social stability.

Social play is one of the most interesting of the ways in which animals appear to maintain familiarity with other group members. Though the young of passerines and penguins and certain other birds seem to play (Nice, 1943), the phenomenon is most conspicuous in social mammals. It is far more conspicuous in the young than in the adult. Young macaque monkeys seem to be as much taken up with play activity as human children. Lambs and other ungulate young play at gamelike activities resembling tag, king-of-the-hill, and other favorites of children. Play-hunting is common in carnivores. Play no doubt serves other functions (Beach, 1945), but it must help in establishing and maintaining learned social affinities. It is noteworthy that play is rare in the adult of most mammals where social relations are already well stabilized. However, in wolves, in which adult activities require the maintenance of a high level of co-operative behavior in hunting, play, especially play-fighting persists among adults. Play may thus be considered one of the techniques of reiterated stimulus exchange by which social animals maintain their familiarity with each other as individuals. It appears to be an important technique by which young social mammals learn their place in the group and develop appropriate in-group feeling. It is often conspicuous in maintaining pair relations in social mammals. In this respect it replaces the kumpan relation commonly serving these functions in lower vertebrates with more stereotyped behaviors.

WILLIAM ETKIN

4

Types of Social Organization
in Birds and Mammals

Introduction

The concept of adaptation implies not only that all the important
characteristics of the organism are suited to the needs of the organism
in relation to its environment but also that these characteristics fit
with each other to make for efficiently operating wholes. For example, the eye of a given vertebrate not only must be an efficient
organ for accomplishing visual needs but it must be efficient in
precisely those ways which the activities of that organism require
and not in other ways or in higher degree than required. The focusing
capacities of the eye vary enormously in different vertebrates, according to the different ecological requirements of the organism. Fish
living in water necessarily have little range of vision and little need
for much change in focus in their eyes. Correspondingly, we find the
focusing mechanism in fish generally poorly developed or absent.
Ground-living forest dwellers require, and have, greater capacity for
accommodation; arboreal and plains-living animals have still more,
and certain birds have the most extensive apparatus. Not only does
the range of accommodation vary appropriately in these differing
animals, but in each type the speed of focusing varies in correlation
with the rate of locomotion of the animal. Although some mammals
have keen vision and considerable depth of focus, their speed of
focusing is slow in comparison with the focusing mechanisms of
hawks and other birds of prey. These have accommodation mechanisms which employ rapidly contracting striated muscle, in contrast

to the slow acting smooth muscle found in the mammalian eye. The hawk must necessarily keep its prey in accurate focus until the moment of contact while pouncing upon it at a speed of some two hundred miles per hour.

In a similar way the behavior system of an organism must not only be efficient but must correlate with the entire life of the organism. Survival value depends upon the over-all balance of many factors. The integration of behaviors is as significant as the usefulness of the behavior considered in isolation.

It must not be thought, however, that integration implies that all the forms of behavior are necessarily harmonious. They may, in fact, clash and, by opposing, keep each other in check. The conflicts which arise in territorial behavior and the means of their resolution illustrate this. Defense of the territory necessitates aggression against other members of the species. If carried too far, this interferes with acceptance of the mate. The level of aggression that is attained in a given species depends upon the balance of such factors as the role of territoriality, extent of sexual dimorphism, the nature of the courtship, role of the male in parental care, population pressure, danger from predators, etc. Tinbergen provided another example by pointing to the limitations imposed upon the development of conspicuousness in signaling devices by the dangers of exposing the animals to predators. To understand the integration of factors in social organization, we must think of each characteristic of behavior in quantitative as well as qualitative terms. In this chapter we propose to describe how the various elements of the behavior systems interrelate in each of several selected animals to produce a type of social organization that constitutes an effective mode of environmental adaptation.

Animal life is nothing if not variable. For almost any general statement that can be made about a group, some exception or apparent exception can be found. We can deal here only with what we regard as the generality of the interrelations representative of the principle of correlation and make little or no attempt to consider the variants.

Since this chapter reviews by animal type behaviors discussed previously in other contexts, it is neither necessary nor practical to specify the references again. With few exceptions, therefore, no direct reference to the literature is given for these items. The most important studies upon which the discussion of each type is based are listed according to types at the end of the chapter.

Birds

PERCHING OR SONG BIRDS

Song birds are highly mobile, diurnal animals, dependent upon vision and hearing for sensory guidance, feeding on small animals (insects, etc.) or plant products, and subject to predation from the air (hawks) and, to a lesser extent, from ground-living animals. Their behavioral system in the reproductive phase tends to be organized around seasonally defended territories limited to one mated pair. In most species male and female are alike (monomorphic). Agonistic display usually involves stereotyped sounds and movement rather than structural characteristics. Courtship is largely limited to pre-nuptial displays and is commonly of aggressive nature. Parental duties are shared more or less equally between male and female, or at least the male is helpful in raising the young. The family unit is disrupted at the end of the reproductive season, when the young become entirely independent of their parents. In both resident and migratory species flocks are formed; often different age classes or sexes flock separately.

A little reflection will show that this catalogue of behavioral items is no mere random assortment but that the factors are interrelated and interdependent. Since the male establishes his territory before mating, the pairing-up process involves the violation of the male's territory by the female. Hence, we expect male aggression to be present in courtship. If pairing is to be successful, the aggressive responses of the male must be countered by some characteristic of the female. Morphological releasers identifying the sexes would serve this purpose, and, indeed, such are found in the red-winged blackbird and flicker. Dimorphism has, however, the disadvantage of rendering the individual more conspicuous and, therefore, subject to predation. Most song birds do not exhibit dimorphism but depend upon behavioral responses, of which the passive acceptance by the female of aggression by the male is the most commonplace. Thus the characteristic courtship of song birds emerges: aggressive actions by the male and passivity by the female—as described for the song sparrow in Chapter 2. Once pairing up has been effected, courtship practically ceases, since the pairing bond is reinforced by the fact that both animals are confined to the territory. Territoriality of the song bird type also limits competition among males to the period of establishment of the territory. Aggressive display in these circumstances is sufficiently accomplished by song, posturing, and minor structural characteristics, and there is a general absence of elements tending to make the males permanently conspicuous. This permits males as well as females to assume protective colorations consistent with the important role that both parents of these species play in

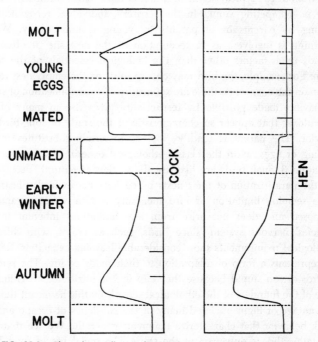

FIG. 10.1. The robin redbreast hen and cock both sing prominently in their individual territories in autumn. Thereafter the female (*right graph*) shows little song display, being especially quiet during the breeding period. Males, on the other hand, show considerable singing especially when unmated in spring. Immediately after mating and at molt they are quiet (*left graph*). (After D. Lack, *The Life of the Robin*, 1943.)

parental care. Even the danger of such an apparent exception as the red breast of robins is minimized by the fact that it is on the under side and only becomes conspicuous when exposed by the specialized postures associated with territorial defense (Fig. 10.1). The breakup of the family bonds when the young reach full flight capacities is correlated with the preparation for the fall migration. This requires a new kind of socialization without territoriality. Since feeding in flocks on widely distributed foodstuffs is non-competitive, little aggressive behavior is shown at this time.

Among song birds responses governed by simple releaser mechanisms predominate, with little dependence upon learning in evidence. For example, the responses between parents and young are largely controlled by typical kumpan relationships. Since ordinarily there are no competing young in the territory, individual recognition of young by parents or of parents by young is unnecessary. Where learning is involved, as in recognition of the nest site, it often depends upon factors other than the biological essentials of the situation. For example, the bird may choose to sit on the empty nest rather than on the eggs when these are set nearby. This simplification of social behaviors made possible by territoriality gives rise to many of the paradoxes that appear so characteristic of the study of song-bird behavior. The parent is readily "fooled" by crude substitutes for its young or eggs, as in the classic ethological experiments discussed in Chapter 3. Such birds are likewise the victims of natural "deceivers," as the parasitization of their nests by cuckoos or cowbirds illustrates. The seeming limitation of bird mentality is clearly not organically imposed and does not arise from any limitations inherent in the species' nervous system, since birds, such as crows, with different ecological requirements show considerable learning capacities. Rather, it represents a form of adaptation to their mode of life. The remark, "Birds are so stupid because they can fly," is attributed to Heinroth, one of the founders of the ethological school. By this he meant that the advantage of flight obviated many of the subtleties of escape and attack behavior that characterize mammals generally. We see that this, though valid, is only part of the story. The social life of song birds is so simplified by their type of reproductive territoriality that there is little need for complex learning activities. The evolutionary process does not call forth any greater or more complex mechanisms than those that suffice to achieve survival under the particular ecological situation.

There are many contradictory and conflicting forces acting upon any organism, and the behavior system achieved in its evolution represents a compromise or resolution of these forces—or perhaps, better said, a dynamic equilibrium between them. For example, the territoriality of song birds would seem to create a selective pressure toward dimorphism, emphasizing as it does aggressive and display potential in the males. The monogamous mating which is favored by the same territoriality, however, creates a selective pressure away from dimorphism, since conspicuousness in the male would subject him to predation and undermine his usefulness in the care of young. In general song birds tend to be monogamous (during each reproductive period) and monomorphic. Sometimes the balance of factors leans toward some conspicuousness, as in the English robins where both have the red breast and both parents defend the territory after

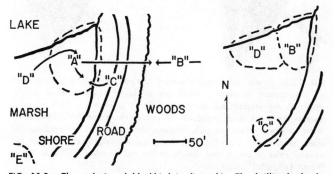

FIG. 10.2. The red-winged blackbird is dimorphic. The brilliantly feathered male holds a territory in which several females may become established. Each female has her own sub-territory around her nest, which she defends against other females. Males attract females to their own territories by courting behavior, but established females often interfere, attempting to drive the new female away. In such cases the male may or may not attack his original mate. The diagram above (left) shows the events in an area in Wisconsin from March 20 to April 21, 1952. Male A held male B far back in the woods. On March 30, male C arrived and challenged A. While A was fighting and driving off male C, male D quietly took over a portion of his territory. Later male A disappeared from the scene and his territory was then taken over by males D and B. Male C, after being repulsed by A, established himself in a territory nearer the shore (right). (After R. Nero, Wilson Bull., 68 [1956].)

pairing. In the red-winged blackbird, there is considerable dimorphism in the black color and red epaulets of the male, in contrast to the mottled browns of the female. This bird has achieved this different balance favoring dimorphism, since the male often has several females in his territory and himself plays little role in parental care (Fig. 10.2, Fig. 10.3). Thus, to appreciate any one characteristic, as aggressiveness toward others of the species or dimorphism, we must see it against the background of the entire mode of life of the organism.

FIG. 10.3. In the red-winged blackbird, the male defends a territory with a "song-spread" display in which the bright-red covert feathers at the shoulders are erected and sometimes vibrate, producing a flash of color against the black background of the rest of the body (*left*). Several females may nest within one male territory, each defending her area against other females with much the same "song spread" (*right*), even though her general color is cryptic and she lacks the red epaulets. (After R. Nero, *Wilson Bull.*, 68 [1956].)

MARINE BIRDS

Marine birds feed along the shore (gulls) or by fishing in the open sea (gannets). Their feeding ranges are therefore very extensive. Feeding is generally in flocks, which, as pointed out in Chapter 1 is more effective than non-social feeding. Breeding activity, on the other hand, is necessarily restricted to fixed shore positions and generally is characterized by the development of small breeding territories within large breeding colonies. Both parents range widely to feed, and both return to the breeding territory to take part in parental activities (Fig. 10.4). After the breeding season the animals may range far from the breeding grounds, generally in flock formations. Since these birds are usually large and strong, the predators from which they suffer are primarily those that attack the eggs and young.

The behavioral correlates of this mode of life include the following: Pairing up usually takes place in the flock outside of the territory (Fig. 10.5). The pair then takes up a very small nesting territory within the colony and both parents defend it vigorously. The parents share nearly equally in the care of the young. Post-nuptial courtship in the form of nest-relief ceremonies is often elaborately developed

(Fig. 10.6). Monomorphism prevails, male and female being so much alike as to be difficult to distinguish. Though initial parental reactions to the nest site, eggs, and young are largely governed by simple releaser mechanisms, parents soon learn to recognize their own young, and, of course, the members of a mated pair show extraordinary ability to recognize the individuality of their mates among the hundreds of highly similar animals all around. Loss of parental care

3

FIG. 10.4. Gannets in a breeding colony each occupy a small breeding territory which is vigorously defended against strangers. In the center, the nest forms somewhat of a mound of sticks, sea weed, etc., in which a single egg is found. Note the regularity of the spacing of the birds. It would appear to require a considerable feat of memory for a bird to locate its own nest, yet this is regularly accomplished with great precision. The mates then indulge in an elaborate nest-relief ceremony before the returning individual is allowed to take over the task of incubation. (Drawn from photograph.)

usually occurs abruptly at the appropriate stage of the development of the young and is often accompanied by definite rejection behavior which compels the young to undertake feeding flights on their own.

The principal difference in the mode of life that we note here in comparison to the song-bird pattern is the separation of the feeding and breeding situations. The persistence of courtship (nest-relief ceremonies) throughout the period of parental care appears to be adaptive to this situation, since the parents must leave the breeding territory to feed and return to the confusion of the breeding ground.

Correlated with this is the highly developed learning ability shown in recognition of mates, which is obviously necessary if the birds are to go from and come to their territories freely. Vigorous defense of the breeding territories is essential for protection against nest robbing. The elaborate nest-relief ceremonies appear to permit the suppression of the highly developed territory-defense reactions when the mate returns. The monomorphic characteristic, as in song birds, correlates with the similarity of the parents in relation to care of the young. Here, however, territorial defense is at closer quarters, and morphological display characteristics are even less evident than in song birds; such elaborations as song and flight displays are notably absent, as would be expected under the conditions of crowding usually obtaining at the breeding site. The fact that pairing takes place before territory establishment is correlated with the small size of the

FIG. 10.5. Food-begging is a common courtship pattern in birds. The male gull (*right*—note swollen neck indicating regurgitation) often feeds the female during courtship. At this time she begs food from him with the same begging movements the young animal shows to its parents (*left*). (After N. Tinbergen, *The Herring Gull's World*, 1953.)

territory. Its size is usually such that the aggressive courtship of song birds would be impossible in it.

As has been discussed in the section on courtship and by others, social stimuli sometimes stimulate the endocrine system through the brain-pituitary connection. The dense clustering of sea birds in their breeding colonies can be understood as an example of the effects of such stimuli, for it has been found that the correlation of activities among different members of a colony are better achieved in large colonies than in smaller, presumably because of the influence of the social stimuli provided by neighbors to one another. This, in turn, makes for better reproductive success of the colony as a whole. For example, if the young of one pair begin wandering from their nest sites while neighboring pairs are still vigorously defending territory, they are likely to be killed. The importance and delicacy with which such social factors operate is seen in instances where, as a result of outside interferences, the social interactions among colony members

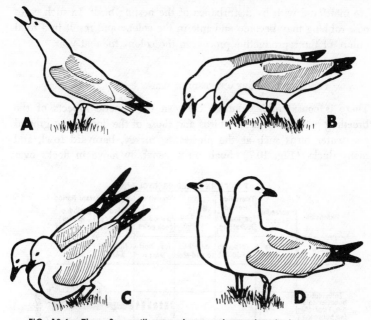

FIG. 10.6. These figures illustrate the mutual courtship displays in gulls. Oblique postures of a male herring gull (A) at his territory shows aggressive motivation, since it is part of the attack posture. This signal repels males but attracts females much as the singing of a territorial song sparrow. When the female arrives, the mutual courtship that follows shows many displays that indicate the mixed motivation of aggression, fear, and sexual attraction. Mutual mew calling (B) indicates aggression since it is often given in frankly conflict situations. However, in the conflict, individuals face each other with their threats instead of assuming parallel positions close to each other as in courtship. Choking (C) likewise indicates aggression modified by the tendency to stay. In the kittiwake, choking is given only in the nest situation and thus appears to include the parental motivation. The movement itself appears as the intention movement ritualized from the regurgitation activity in feeding young or mate. Facing away (D) in the black-headed gull appears as aggression inhibited by fear. It appears to be derived from the intention movement of turning to flee. As a signal of low level of aggression it serves as an appeasement sign which inhibits the aggression of the mate. Thus the courtship, pre- and post-nuptial, of the gulls shows displays indicating different mixtures of motivations and appearing to be signals "correctly interpreted" by the interacting gulls. (After N. Tinbergen, *Behaviour*, 15 [1959].)

are interfered with by disturbance of the nesting birds. In such cases nest robbing may become endemic in the colony and result in a total failure of the reproductive process in the colony for that year.

FLOCK-LIVING SURFACE-FEEDING BIRDS

There is considerable contrast between the general aspects of the breeding patterns of marine birds and those of the flock-living ground and water birds such as the pheasants, turkey, barnyard fowl, and many ducks (Fig. 10.7). Such birds generally move in flocks over

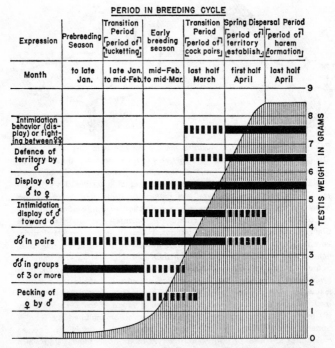

FIG. 10.7. The sequence of behavioral characteristics in pheasants during the year is shown here. During the winter the males tend to stay in unisexual flocks and show aggression toward females. At this time their testis (*shaded area*) are small (*two lowest spaces*). As spring approaches the males separate, show aggression toward other males, and display to females. In the April breeding season the males defend the territory, and the females show aggression toward each other. The testes have reached maximal development at this time. (After N. E. Collias and Taber, *The Condor*, 1951.)

large areas, feeding chiefly off the surface. The young are necessarily precocious and follow the mother and feed for themselves early, often when only a few hours old. Females and young are protectively colored, but the adult males are among the most conspicuously and gorgeously attired of all birds.

These birds also differ behaviorally from marine birds to an extreme degree. There is no reproductive territory established by the males though special mating territories may be established in some species. In many species each male forms a harem of females which he dominates and from which he excludes other males. The female usually defends the nest site. The dominance hierarchy in the flock is clear-cut, linear, and absolute in form, aggressiveness being especially conspicuous among males. Parental care is typically left almost exclusively to the females, although dominance may be modified in the direction of males calling females to food (chickens). Males act as the defenders of the flock. Responses to predators depend chiefly upon innate releasers of instinctual escape reactions. Learning ability is conspicuously developed in respect to individual recognition of members of the group—as is necessary for hierarchy. Food discrimination is also readily learned, as would be expected in general foragers (chickens). The learning of parent-young relation furnishes the paradigm of imprinting (chickens, ducks). The adaptive nature of this constellation of behavioral and structural characteristics for surface-living birds is readily seen. Flock-living is protective in preventing surprise by predators and as such is of supreme importance for surface-living, diurnal animals. This and the need for a wide foraging range preclude the limited territoriality of the song bird type. Flock-living gives prominence to the development of dominance hierarchy; and this, operating in the sphere of reproduction, places a high premium upon male dominance, since only the dominant males will succeed in breeding (Fig. 10.8). In such flocks, too, the success of reproductive behavior depends upon female responsiveness during estrus to the dominant male. These factors lead to the extreme aggressiveness and elaboration of display characteristics of the males, as noted above. Furthermore, the precocial young require no parental care from the male and only a limited amount of protectiveness from the female. The male role is thus limited largely to insemination of the females, and the expendability of the males in the harem situation, where only a few males are required for reproduction of the group, removes much of the restraint upon the development of con-

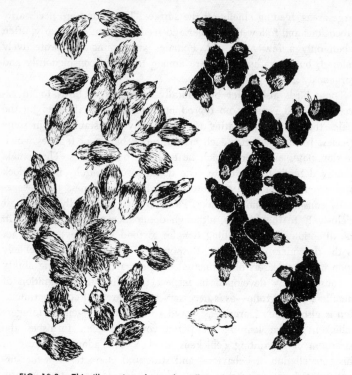

FIG. 10.8. This illustration shows the effect of dominance on breeding success of roosters. Three roosters of different strains were penned with white hens. The offspng of the Rhode Island Red are recognizable by their dark color, those of the barred Plymouth Rock are lighter and the one chick fathered by the white leghorn is white. Since the sperm of each of these varieties were shown to be equally fertile, the difference in the number of offspring shows the difference in access to the females of the roosters according to dominance order. (Redrawn after A. M. Guhl, *Scientific American*, 1960.)

spicuousness in the males which may be operative in other situations. Feathers and other ornaments have been developed to an extreme in the males of such birds (Fig. 10.9). Lek formation in some members of this group is a type of behavioral organization possible to such non-parental males and appears to take place in forms which are widely dispersed in feeding. It thus enables the females, when ready for mating, to find the males more easily.

In these animals the different roles of learning and instinct can likewise be recognized as adaptive. These are the animals in which the classic examples of absolute dominance have been so extensively studied. It can be seen that the ability to learn individual recognition is the basis for flock organization. The size of the flock in nature appears to be roughly co-ordinate with the extent of this learning ability. The closed nature of the flock formation and the extreme xenophobia shown may likewise be understood in the light of the control of aggression possible only in the hierarchically organized flock. Again the phenomenon of imprinting in parent-young relation is readily studied among such animals and is clearly part of the system of behavior that makes their survival as mobile flocks possible. On the other hand, the predator relations of these birds is much like that of song birds, and the signal system that activates their hiding-from-predator response also depends upon innate releasers.

The conclusion of our consideration of three different types of social organization among birds is that in all fundamentals, their

NORMAL MALE   NORMAL FEMALE

CASTRATED TYPE   CASTRATE WITH   CASTRATE WITH
                            OVARY   TESTIS

FIG. 10.9. The complex interrelations of hormones and the secondary sex characteristics which determine sexual dimorphism are illustrated in domestic fowl. The comb and wattles of the head are under the influence of male hormone. This hormone is produced in high concentration by the testes and in low concentration by the ovary. Hence the male has large head furnishings, the female small, and the castrate none. Male plumage develops when the individual matures but it does so only in the absence of female hormone. We find, then, that castrates or castrates with grafts of testes have male plumage but, with ovarian grafts, have female plumage.

behavioral factors—such as aggressiveness, hierarchy, territoriality, courtship, parental care, dimorphism, innate or learned communication—are all part of a co-ordinate complex which, in its entirety, is adaptive to the particular animal's ecology.

Social Organizations among Mammals

THE RAT

The wild brown rat, of which the common laboratory animal is an albino-mutant form, makes its living as a scavenger, being active at night and depending for protection upon its ability to find cover quickly. Its territoriality is essentially of the home-range type, accompanied by an intense exploratory drive. No typical defended territorial reactions are shown; however, each group tends to remain separate from others, with an uninhabited no man's land between home-range territories. Though dominance patterning can be brought out to a limited extent in artificial competitions, it is neither markedly expressed nor quickly established. Such hierarchy as is shown under experimental conditions seems based on individual habits of aggression, with little evidence of recognition of individuals as such. Most important, there is no competition between males for an estrus female, for one rat (in laboratory albinos) does not exclude another from access to an estrus female. There is slight sexual dimorphism, with males somewhat larger than females but otherwise very similar. They breed rapidly, being polyestrus during the major part of the year and, under artificial conditions, throughout the year. The young grow rapidly, reaching sexual maturity in about two months. Parental care is given only by the female and is restricted to a lactation period of about twenty-one days. Litter mates do not maintain continued association so that no "pack" organization results, but the individuals tend to remain in the same colony.

The central area of adaptation in this behavioral system appears to be that of rapid population expansion. This enables them to build up a large population to take advantage of temporarily favorable conditions, such as are afforded by seasonal food supplies. The absence of defended territory is adaptive here, avoiding as it does restrictions upon maximal reproduction. The physiological pattern of reproduction likewise favors rapid population expansion. Since gestation and lactation are of about equal length, the female nurses one litter while carrying the next. A few hours after giving birth, she again comes into heat, thus starting the next pregnancy. This reproductive pattern sometimes gets out of hand and gives rise to the familiar phenomena of mouse and rat plagues. Their intense territory-exploration drive, which is, of course, the basis of their usefulness to psychologists in maze-learning experiments, serves as part of their protective behavior. As mentioned in Chapter 3, rats kept under semi-natural conditions come to know their home territory so thoroughly they can instantly find the best and nearest shelter whenever disturbed, no matter what part of their range they may be in at the time. The highest learning capacities of rats are thus related to home range rather than to hierarchy, sex, or feeding. Their mating and parental activities seem to be largely governed by innate stimulus-response relations and stereotyped motor patterns, as might be expected from the relative simplicity of their reproductive life.

THE RED DEER

The red-deer herd is matriarchal in organization. It is derived from the association of the young with their mothers, and thus it develops from the maternal family. Since the animals are large and relatively long-lived, three or more generations are commonly represented in the herd. The young all associate with their mother for the first three years; the males then maturing sufficiently to become sexually active at rut. They lose their association with the herd, whereas the females remain with the maternal herd. During the year the animals move through their home-range territory in a schedule dictated by seasonal conditions—the highlands in the summer and the protected valleys in the winter. The mature males live in loosely organized herds separate from the females. The latter live in herds which include the young and which are well organized. Dominance and leadership are exercised by the older females, one of which often exerts clear-cut leadership, guiding the migration, urging laggards over streams, and in other ways directing the herd's activities. The older females are constantly on the alert for signs of danger, from predators and from adverse weather changes. It is clear that they determine the movements of, protect, and guide the herd through the regular paths of migration used year after year (Fig. 10.10). In the fall rut season the male herds break up and the individuals scatter widely. They invade the female territories, round up the hinds, and keep them in

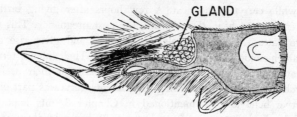

FIG. 10.10. This is a longitudinal section of the front foot of a mule deer doe showing location of interdigital gland. This gland produces secretion that lays down a scent trail by which the animals can follow each other's paths. The secretion is particularly heavy in young fawns, probably enabling the mother to trail them. Males rub their legs together thus distributing the scent of glands on the tarsal regions. They also trail the does by the scent and appear to distinguish estrus does by the smell of the urine. This is important, for in this species the males do not herd the females into harems but pursue them individually. (After J. M. Linsdale and P. Q. Tomich, *A Herd of Mule Deer*, 1953.)

harems. These males display aggressively to other males which try to approach and, as necessary, fight to exclude other males from their harem areas. As the females come into heat, they are serviced by the harem master only. This constant activity, sexual and defensive, maintained day and night by the stags usually exhausts them after about a week, and they are then forced by fresh animals to retire to the uplands, leaving their harems to the newcomers. Foals are born in the early spring, stay close to their mothers, and achieve substantial, but incomplete, growth before the first winter.

Among the behavioral correlates of this mode of life, we may note that the males show little socialization, separating from the female herds when young and showing only the loosest herd structure thereafter. In correlation with this is the strongly developed aggressive activity of the rut and the polygynous mating pattern. The role of the male is merely that of insemination, and his aggressive actions and the extreme development of aggressive potential are consistent with the dispensability of the individual male for parental functions and the polygynous structure of the society. As is to be expected in this situation, the females show little courtship activity. They tolerate the harem master but do not co-operate in his territorial-defense interests. They do, however, show clear-cut dominance hierarchy within their own herds. This hierarchy is based upon learned relations between mother and child during development. In the adult stage no aggressive action is necessary to maintain the leadership precedence. The young females are attentive to the movements and actions of the older females and appear by this attentiveness to learn the "traditions" of the herd with respect to the seasonal migratory paths, the interpretation of weather signs, and sentry activities. In contrast the male yearlings (according to Fraser Darling) are not so attentive to the actions of other herd members and do not acquire the store of knowledge of the herd. In correlation with their highly developed dimorphism in respect to aggressive potential and their pugnacity during mating, the males play no role in parental behaviors, neither protecting, assisting nor even paying any attention to the young. Except when maintaining the harem, the males show no aggressiveness toward one another. In contrast to the female herd, no hierarchy, leadership, or co-operative behavior is evident in the male herd. The intricate relations of these animals, especially the females, to one another and to the changeable weather conditions upon which their survival depends correlates with the lack of stereotyped behavior and the prominence of learning in their activities.

THE WOLF

Wolves are pack-hunting carnivores that bring down large prey, such as deer, by co-operative action in hunting. They not only attack in a pack but carry through such pack stratagems as tiring out their prey by spelling each other off in keeping up the pursuit. The pack appears to be derived from the family unit, consisting of father and mother and offspring of both sexes and of several successive years. Apparently on occasion several such family packs may join forces, at least temporarily, to form large packs. Whether the subgroups in a large pack are blood relatives, i.e., have common familial experiences forming the basis of their tolerance for one another, or not is not known. A breeding pair generally has several dens—either natural caves or excavated holes in banks. Male and female remain together apparently for life. Adult non-breeding animals, presumably the older offspring of the senior pair, have been seen to remain associated with the pair and their young litter and to share in all activities. When burdened by pregnancy or newborn young, the female does not hunt with the pack but is fed by the male or other members of the pack who bring back part of the kill to her. After the pups are born in the spring, food is brought back to the den by all the adult wolves, sometimes in chunks and sometimes by regurgitation. The pack appears to be a closed group, since strangers have been seen to be driven away from the group. But on other occasions, new animals were seen to join a pack. Possibly these were former members of the group. There is much play—mouthing, licking, chasing, and play-fighting—among all members of a pack at all times. Friendly greetings with tail wagging and howling are frequently exchanged between members of the pack, particularly on special occasions, as when the pack gathers together before going off on a hunt.

We note here a number of behavioral correlates to an ecology centered around pack-hunting. The female with a new litter and possibly also one late in pregnancy is unable to hunt with the pack. Co-operative behavior, marked by the bringing back of food to the den by the other adults, is adaptive in this situation. Such behavior is only one of the more dramatic examples of the co-operative behavior that marks the pack-hunting mode of life and without which this way of life would be impossible. The inter-individual behaviors of the members of the pack are marked by much social communication by play in various forms, not only among the young, but also among the adults. This would appear to be a means of constant reinforcement of the individual social interdependencies. Of course, the members of a litter, as they grow up together, engage to an even more marked degree in this social play. As in many other carnivores the young learn hunting techniques from the adults; but in this case the close socialization is not restricted, as in cats, to the mother-young relation but extends to all members of the group. Socialization by various forms of play with other members of the litter of the same generation is especially marked and appears to be the primary bond between members of the pack. There is little stereotyped behavior to suggest kumpan relations.

Among mammals the wolf type of socialization is exceptional, though not unique, in including a dominant male (father) and other non-breeding males in a co-operative group. Such a tight social group, including all members of the biological family, may be called an integrated family. Dominance hierarchy is clearly present, but its expression involves a minimum of agonistic behavior since it forms during development of the young within the family relation and remains stable throughout life. The absence of fighting is not the result of an absence of dominance, but rather, a result of the stability of the hierarchy. The extraordinary phenomenon of adults remaining within the pack without showing breeding activity, as appears to be the case in wolves, may receive its explanation from this strongly developed dominance hierarchy; for, as discussed in Chapter 2, experiment has shown a tendency in dogs and other animals for mating behavior of males to be suppressed in the presence of a dominant male (psychological castration). The system of communication between individuals within the pack is quite elaborate, as might be

SELF-CONFIDENCE   **THREAT**   **UNCONCERN**

UNCERTAINTY   FRIENDLY SUBMISSION   **SUBORDINATION**

FIG. 10.11. By variation in carriage and facial expression wolves communicate their social reactions to each other. In an extensive study of wolves under zoo conditions, Schenkel learned to interpret this signaling system. His inferences of the significance of some tail positions is shown above. It will be readily recognized that much of this signaling system is common to domestic dogs. (After Schenkel, *Behavior,* 1 [1948].)

expected. It involves much signaling by body, facial, and tail movements, posturing, and some vocal expression (Fig. 10.11).

MACAQUE MONKEYS

As studied under semi-natural conditions, the Indian or rhesus monkeys and the Japanese macaques (also baboons) live in clans of some 20 to 150 animals. Their habitat is forest or semi-open country. They feed in trees or on the ground by food gathering, eating mostly plant products but also insects, bird eggs, and small animals when available. Except for nursing infants, each animal forages for itself. Usually several males are found in each clan. These males are supremely dominant within the total group and are themselves arranged in a strict linear hierarchy. Other mature males are excluded and form associated bachelor groups. The dominant males show little or no co-operative behavior toward the females, though they may take important positions when the clan is in danger. In the Japanese species, the males are reported to be somewhat more social to females and young, even playing with and caring for young to a slight extent in some groups.

The group territory is defended by the dominant males. Territory fighting is vicious, and the fighting capacity of the top males is a principal factor in determining territory size. Females in heat approach closely to the males, usually facing some adverse aggression in their initial approaches. Often a female passes from subordinate male at the beginning of estrus to the top dominant at the height of her estrus and then to a subordinate again. The young from birth associate closely with the mother, who nurses, cares for, and supervises them. Socialization depends upon these female-young contacts for its initiation but is maintained by widespread grooming activity among all members of the group and by play in youngsters. Facial expressions, postural changes, and sounds are used as signals. The necessary learning includes individual recognition, complex pathways through trees, the variety of and methods of extracting food materials, and possibly the signal system. Dominance and sexuality are closely interlocked. The sexual presentation posture is assumed as an expression of subordination and serves to halt aggression on the part of the dominant animal by channeling such aggression into mounting behavior. Mounting is thus a characteristic behavior of the dominant, as the presentation posture is of the subordinate, irrespective of the sex of the individuals involved. These behaviors are thus not strictly sexual in these animals.

The social organization of the macaque monkey is seen to emphasize the evolutionary pressure toward male aggressiveness. The dominant males exclude subordinates from breeding, and most of the offspring are thus probably fathered by the dominant males of the group. The aggressive potential of the males is further emphasized by the group territorial system, which makes group success in holding desirable territory dependent upon such potential. The exaggeration of the canine teeth, the large body size and power, and the aggressive disposition of the males is the result to be expected from these evolutionary pressures. The importance of aggressive activity in this social organization is further emphasized by the existence of bachelor males ever on the alert to invade the group.

It is difficult to see in what way intelligence and learning are favored in this ecology. It is true that the attainment of goals by roundabout paths demanded from animals with wide-ranging feeding in trees places a premium upon the type of "problem-solving" learning in which monkeys excel. The precision of binocular visual discrimination and judgment and the use of the hands in manipulation and feeding likewise favor a kind of intelligence. However, their feeding activities seem much less demanding of intelligence than the co-operative-hunting techniques of wolves.

A somewhat broader picture of the role of intelligence is emerging from the recent studies on Japanese monkeys. Food preferences and selection among the hundreds of available substances are learned by the young from other members of the group, usually the mothers. A new preference or way of handling food (i.e., washing) acquired by one adventuresome individual has been seen to spread through the group and become a culturally transmitted trait, distinctive of the group. Differences between clans in mounting activity and in the co-operativeness of males with the young are also reported and are probably culturally transmitted. Social factors, such as relationships of females to males and of young to females, are reported to influence position in the dominance hierarchy in subtle ways. The importance of these cultural factors suggests that a high level of intelligence is important in monkeys, primarily in relation to the complexity of their social life.

HOWLING MONKEYS

Howling monkeys of the tropical rain forest live in groups much like those of macaques, except that aggressive behaviors are much less in evidence. The males show little evidence of dominance and do not even compete for the estrus females, nor do they exclude younger males from the group. The males are not indifferent to the young, although their co-operative activity is rather limited. For example, they call to attract attention to young which have gotten into difficulty by falling, although they leave the work of rescue to the females. They are not, however, aggressively negative toward young, as rhesus males often are. The males are most active in seeking pathways through the trees of the dense jungle in which they live. They direct the movements of the group by vocal signals, calling on the females and young to follow. Among the most unusual activities of the males are their howling sessions, which regularly take place in the early morning and produce a tremendous volume of sound. This is probably a signal that warns different clans of the presence of others and serves to keep the groups apart, despite the density of the plant life hiding one from the other. Several clans may cover the same general territory in their home ranges. If two clans come into close range during the day's movements, a competitive howling session determines which group yields to the other.

Compared with that of the macaque, this social system is seen to place much less emphasis upon aggressive behavior. In correlation with this, the dimorphism of the sexes is not nearly so marked as in macaques. It is possible that the permanent positions of the animals —the high tree tops of the tropical rain forest with its very dense plant growth—is basically responsible for the minimizing of fighting and the substitution of howling as a competitive mechanism. Obviously, fighting in their habitat is both dangerous and ineffective. Correlated with low aggression and lack of dimorphism is the more co-operative behavior of the males in relation to females and young. Whereas the macaque male generally ignores females and young when not actually aggressive toward them, the howling-monkey male leads or brings up the rear of the group in a protective manner. He rushes to the defense of members of the group, a behavior distinct from the defense of territory by the macaque, which ignores individuals. Altogether, the differences in behavior of the two monkeys seem adaptive to the different environmental situations of the two groups and the resulting differences in the selective forces playing upon them.

General Principles of Behavioral Correlation

We may now attempt to summarize briefly some of the main areas of correlation that have emerged from our consideration of a few specific types of social organization in the animal world.

DOMINANCE

Dominance hierarchy is best developed where situations of competition exist between individuals within the group, most often with respect to reproduction. Behaviors correlated with hierarchical organization include highly developed individual recognition of in-group members and antagonism to out-group individuals. In-group feeling is generally maintained by learned communicative activity, as in play and grooming, early experience being especially significant in establishing the relationship. With respect to reproduction, hierarchy favors the dominant animals in mating. It correlates with polygyny and harem formation, with dimorphism in terms of male aggressive potential, and with an absence of co-operation by the male in parental duties.

TERRITORIALITY

Home-range territoriality correlates with the needs for security from predation and often is expressed in a highly developed exploratory drive. Defended territoriality generally correlates with the nature of the reproductive pattern. Defense is usually a male function. Group-defended territory is correlated with polygyny and exclusion of young and subordinate males from the group. The small breeding territory of a mated pair, as in gulls, is correlated with bi-parental defense of territory, monomorphism and bi-parental co-operation in raising young. Recognition of individuality in mates and in young (when these become actively mobile) is well developed. Courtship behavior continues during the parental period in the form of nest-relief ceremonies. Reproductive territory of the song-bird type has some similar correlates, including quick learning of territory, courtship which mitigates the aggressiveness of territory defense, and parental co-operation in raising young. The recognition of mate and young, however, is poor; and there is much dependence upon kumpan relations and innate releasers in social activities—including courtship, mating, and parental activities.

REPRODUCTIVE BEHAVIOR AND DIMORPHISM

The type of courtship shown correlates with the biological characteristics of the reproductive pattern as seen in the differences between birds and mammals. The types of reproductive behavior also depend upon (*a*) differences in the roles played by the two parents, (*b*) the type of territorial defense system, and (*c*) the kind of dominance hierarchy. Dimorphism in the form either of aggressive potential or display potential interferes with co-operation by the male in the parental role and with the survival of such males. Where one parent, usually the male, is free from the requirement of co-operation in parental function, dimorphism is favored, since such males are biologically expendable in the economy of the species. Mammalian reproduction, because of the predominance of the female in relation to the young, tends to have this dimorphic nature, as does that of group-living polygynous birds.

COMMUNICATION AND SOCIALIZATION

The nature of the stimulus-response relations found in a species is highly correlated with its ecology. Sign stimuli are conspicuous where the association of the animals is simple and not subject to too many disturbing stimuli, as in the parent-offspring interrelation in territorial birds. Imprinting characterizes ecologies such as that of ducks and ungulates where individuality of the parent-offspring relation must be maintained under conditions where many similar individuals would confuse responses based upon sign stimuli. Where socialization is of a more complex nature, as in the permanent groups of long-lived mammals, socialization is generally learned on an in-group versus out-group basis. Play and grooming, and other forms of sharing mild stimulation, by providing numerous contacts are the principal techniques for learning socialization (Fig. 10.12).

Application of the Principles of Correlation to Man

INTRODUCTION

The social organization of contemporary man is obviously culturally determined to such an extent that many persons doubt that biological factors play a significant role at all. The philosopher and historian Ortega y Gasset has asserted, "Man has no nature, what he has is history." That is to say, he is what his past experience has made him. However, biologists would incline to seek underlying, biologically determined tendencies even in man and to regard cultural modifica-

FIG. 10.12. The importance of social behavior in the survival of a species is beautifully illustrated in studies of prairie dogs by J. A. King. By living in colonies, they secure protection against predators by their constant watchfulness from their hills and their barking signals. Their grazing habits prevent the growth of tall grasses over a large area. They thereby encourage rapidly growing weeds upon which they feed, which in a way borders on the practice of agriculture. The members of each sub-division of a prairie-dog town constitute a coterie in which each member learns to know and accept the other members of the coterie, while rejecting outsiders. This, of course, is possible because the animals maintain constant social contact with each other by various methods of communication. Upon meeting, members frequently groom each other, play by pawing and climbing over one another (4) and nuzzle each other in an open mouthed kiss. Usually only one adult male (2) and several females with their young (3) constitute a coterie. Each coterie jealously guards its own territory, often less than an acre, and system of burrows. Knowledge of the territory is transmitted from generation to generation. The importance of the sense of smell in their social communication is shown not only in the friendly kissing but also by the fact that strangers turn tail toward each other and expose their anal glands for olfactory exploration (1). (After photographs and description by J. A. King.)

tions as being limited and guided by these. So large a question could not profitably be discussed here. However, as evolutionists, we must suppose that man at one time had a biologically determined social organization, which might be elucidated by the application of the principles of correlation developed above and applied to our knowledge of the paleontology of man. We will here attempt a brief sketch of what might have been the social organization of man at a period before culture was so predominant (protocultural man).

The question of the psychological evolution of early man has attracted a great deal of attention in recent years, as the importance of social behavior in animal life has come to be more and more appreciated. Several books summarizing the results of symposia and a number of journal articles which deal with this subject are listed at the end of this chapter. The discussion here will follow the detailed analysis given in this author's technical papers (Etkin, 1954, 1963a). It centers around the modifications in social behavior involved in the shift from the herbivorous ecology characteristic of old-world monkeys and apes to one in which hunting played a central role. Other points of view expounded in the literature are those of Oakley (1951, 1957, 1962) and Washburn (1960, 1962), who stressed the importance of tool-use and tool-making, and that of Chance (1953, 1962), who emphasized the socio-sexual relations. In accord with the viewpoint adopted in this chapter, the significance of social behavioral factors in the life of early man, particularly in relation to hunting, was noted by Bartholomew and Birdsell (1953), Eiseley (1956), and Imanishi (1961). Washburn and DeVore (1961) have also recognized the importance of hunting to social behavioral evolution in their most recent discussion.

The basic question is the nature of the ecological situation to which protocultural man was adapted. The *Pithecanthropus* fossil material and associated artifacts (Peking and Java man of about one-half million years ago) enable us to visualize fairly clearly the general ecology of that fossil man. He was a hunter of large and medium-sized animals, who used stone tools in the hunt and in other activities. Presumably he also made use of naturally occurring plant products for food. He used fire in the preparation of his food and took advantage of natural shelters, such as caves, for protection. Probably he fashioned rough clothes from animal skins. Thus with a cranial capacity of about nine hundred cubic centimeters, or roughly three-quarters of that of modern man, *Pithecanthropus* had definitely moved into the cultural sphere. In fact, as Weidenreich (1947) has pointed out, his mode of life was not too different from that of primitive hunting cultures of contemporary man.

The next, more lowly organized of fossil types is that of the South African ape men (*Australopithecus*), with a cranial capacity about half that of modern man (450–700 cc.) and not much above that of the larger apes. Our knowledge of its ecology is less clear. The South African paleontologists, particularly Raymond Dart (1960), who found and analyzed much of the fossil material favored the interpretation that this creature, too, lived in good part by hunting game animals, including baboons and antelope. Recent finds of paleoliths and bone industry associated with the *Australopithecines* by Dart (1960) and Leakey (Leakey and Des Bartlett, 1960) strongly support this view. Since these creatures were somewhat smaller than modern man and, like him, devoid of structural weapons (unlike apes, even their canines were not enlarged), hunting could only have been effectively carried out with pack organization and the use of weapons. From the structure of the pelvis, it can be said that the *Australopithecines* were bipedal and quite erect in locomotion. Their hands, therefore, were free for the use of weapons while they ran, in contrast to the situation in modern apes, which depend on their arms in rapid locomotion. We may conclude, therefore, that probably very early in evolution of his distinctive brain, man began the shift from a food-gathering to a pack-hunting ecology.

As we saw in considering the social organization of the wolf, the central condition for a pack-hunting mammal with young that have a considerable period of dependency is an integrated family organization, that is, one in which the male parent is an active co-operating member and one in which the maturing young remain associated with the parents. Such an organization contrasts strongly with that of the food-gathering macaque clan described above. Since man originated from such food-gatherers, in evolving in the direction of pack-hunting, protocultural man must have undergone a social behavioral revolution from something like the macaque type toward the wolf type. With this basic shift in mind, we can analyze the main trends in terms of the principal categories of social behavior used in this book.

SOCIAL DOMINANCE

As in mammals generally, the persistence of the association of young with adults carries with it the continued dominance of the adults over the offspring, even when the latter are grown up—as we saw in the red-deer and wolf organizations. In contrast to the deer, however, the dominant male in hunters must be included as a co-operative member of the permanent social group. Furthermore, no exclusion of the young males from the group as occurs in the macaque could take place if a hunting pack is to be formed. Since the group is co-operative in hunting, dominance must tend toward the mild "leadership" type, such as in the female herd of the red deer, rather than the competitive aggressiveness of macaques. Competition in mating among males must, therefore, be avoided. This can be achieved by the suppression of mating in subordinates in the presence of dominants as we saw it to occur in wolves.

In this way the pack-hunting ecology would be expected to shift the mating system from the macaque type toward the integrated family with the elimination of mating competition among family members. Such a trend, developing in a large-brained primate already capable of considerable cultural regulation of social behavior, could give rise to the cultural incest taboos which are so fundamental a characteristic of human society and so unlike the condition of inbreeding in macaques. The essentials of such a system seem to obtain in wolves. In this view, the origin of incest taboos is to be sought in social behavioral factors rather than in such genetic considerations as the effects of inbreeding. It should also be noted that the recognition of biological paternity or other "blood" relationship is not a necessary condition for the establishment of incest taboos.

TERRITORIALITY

There is a striking resemblance between territoriality as it appears in animal societies and the reactions of human groups to their "native" soil. One would not expect in hunting groups the extremely rigid defense of boundaries that characterizes some societies, such as song-bird pairs. Large predators seem to maintain a somewhat looser territoriality. The territory itself is necessarily large and often mobile, following the migration of game. The joining of small family groups to form a large pack, a well-authenticated phenomenon in wolves and thought also to occur in the gorilla (Kawai and Mizuhara, 1959; Schaller, 1963) indicates that a less rigid territoriality is to be expected in primitive man. Presumably, the groups that join have originated from a common family and have retained some effects of early socialization. A measure of such social acceptance between neighboring groups would furnish the basis for the shifting of individuals between groups. In this way exogamy, a necessary condition for the full operation of the incest taboo, might also be favored in the shift to hunting ecology.

MATING BEHAVIOR

The absence of a definite estrus period is a distinctive feature of the female sex cycle in the human. Although we shall speak of a period of receptivity in the human female, it must be understood that this is not fully comparable to the receptivity of the estrus of infrahuman females, since the latter advertise their condition and actively seek mating. It has been pointed out by many writers that the constancy in sexual receptivity of the human female is one of the characteristics which make for permanent integration of the male into the family life. Our analysis of social behavioral factors enables us to specify ways in which the maintenance of this constant state of receptivity upon the part of the female favors the integration of the male into the family organization. It will be recalled that in mammals generally, sexual congress is limited to the estrus period of the female by her behavioral reactions, whereas the male maintains constant sex interest. The human female is continuously sexually receptive. The sustained sex interest of the partners thus favors the integration of the male (Zuckerman, 1932; Chance, 1962). Furthermore, in the absence of a definite estrus period, the specific stimuli furnished by the usual mammalian female as a sign of estrus (i.e., odors, sexual skin color, etc.) are no longer functional and have dropped out. This must minimize the occasions for male rivalry for the females and favor a generalized socialization process as the primary determinant of sexual activity. As we have seen, play is such a general socializing activity. Sexual play in humans is greatly facilitated by the assumption of the upright posture, which makes ventral copulation possible, replacing the awkward mounting posture found in other primates (Fig. 10.13).

The importance of socialization between mates as a determinant of sexual activity in man in contrast to the direct stimulation of sexual activity by specialized signaling devices at estrus gives further emphasis to the point made above—that sexual activity tends to be governed by dominance and other purely social relations. This, of course, furnishes the basis for the establishment of the cultural taboos that are so prominent a part of human sexuality. The detailed studies on Japanese monkeys (Itani, 1958; Kawai, 1958) reveal a high degree of complexity of social behavior in which consort relations interact closely with the hierarchical structure of the group. Each male, for

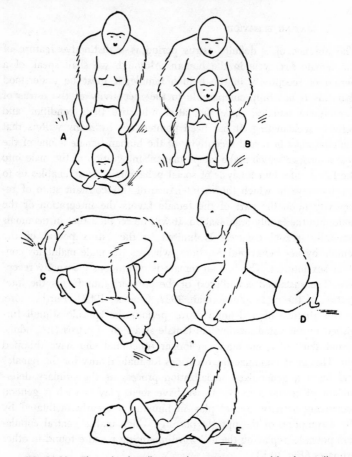

FIG. 10.13. These sketches illustrate the positions assumed by the gorillas in copulation. (Drawings by G. Schaller, *The Mountain Gorilla*, 1963.)

example, tends to have stable consort relations, and the social position of the consort and even of her offspring is a function of the place of the male in the hierarchy, thus adumbrating social class structure. It can readily be appreciated that an evolutionary pressure toward the integrated family would greatly expand such relations and help make the transition to a culturally determined pattern of socio-sexual activity, as we find it in human societies.

PARENTAL CARE AND DIMORPHISM

The principal basis for the evolution of dimorphism in animal societies, as we have seen, is the difference in reproductive roles of the two sexes. This is well illustrated in macaque society by the development of a high degree of aggressive potential, powerful canine teeth, large size and strength, and aggressive disposition in the males, in contrast to the females. Some students have attempted to see in man a behavior structure derivative of this, on the assumption of some sort of evolutionary carry-over from the more general primate conditions. If the general thesis of this book is correct and behavior systems are readily modifiable in evolution by ecological demands, then we should look more to the wolf than to the macaque as a basic model for understanding protocultural man. We found that the male wolf was integrated into the family and provided parental care in correlation with the necessities of the pack-hunting situation, particularly the dependence of the pregnant and nursing female upon other members of the pack. In the human situation, as visualized in this theory, the dependence of the female is further emphasized by the greater size and helplessness of the human infant, factors which would preclude any hunting on the part of the female for most of her active breeding life. This, of course, correlates with the economic role of the female in human hunting societies as food-gatherer and domestic. In marked contrast to the macaque situation, a strong selection pressure must be presumed to have operated toward reducing the aggressiveness of the males within the family situation.

In large predaceous animals group size is generally quite small in contrast to the large groupings of foraging animals such as monkeys. Small group size also reduces the competition between males for females and so limits the trend toward dimorphism in early man. Since only the males could hunt, they must have evolved in the direction of co-operation with all members of the group. They must co-operate with other males in hunting; they must bring back food for the women and children and assume other responsibilities, such as training the boys in the techniques of hunting.

The difference in economic roles visualized for males and females would tend to maintain some differences between the sexes in respect to body build, strength, and endurance, but without encouraging intragroup competition among males. The absence of enlargement in the canines of australopithecine males strongly suggests the absence of competition between males for females. Thus the dimorphism which characterizes modern and ancient men and ape men, namely, moderate differentiation in strength but without strongly developed aggressive potential, is here visualized as adaptive in the ecology of protocultural man. It fits the division of labor between the sexes better than would a sexually competitive pattern among males such as exists in macaque society.

PROLONGATION OF DEPENDENCY

The third distinctively human characteristic in reproduction to be accounted for is the extreme lengthening of the period of growth and learning. In primitive societies the human infant nurses until at least two to three years of age; the monkey baby, for only two or three months; and the chimpanzee, for less than a year. Puberty is correspondingly delayed. As has been repeatedly pointed out, the prolongation of the period of dependency is an essential adaptation of the human condition, making for the educability of the child and therefore for human culture. But what we must also realize is that this could only arise as a continuing evolutionary process *pari passu* with the development of the integrated family ecology; for, unless the male "breadwinner" were incorporated into the family, the burden on the female of having several dependent young would be unworkable. It should be recognized that in an ecology such as that of the macaque, where the young are a burden upon the female only and where she must forage for herself and baby, there is a strong selection pressure for rapid development of the young, just as there is in birds generally (see Chapter 2). The condition of slower development in man may therefore be viewed as the result, in part, of a relaxation of the extreme pressure for rapid development as well as a positive pressure for a longer learning period. This means, then, that the prolongation of the human period of immaturity may have started independently of its value in a cultural system, but, once started, it became accelerated by the fact that within the integrated family situation it permitted the further advantage of cultural transmission.

The final category of social behavior to be analyzed is that of socialization; this has already been touched upon above. Of the factors which may be subsumed under this heading, we should consider three at this point: language, play, and cultural transmission, as presenting distinctively human problems.

LANGUAGE

The central aspect of the problem of language, from the biological viewpoint may be stated as follows: What selective forces led to the development of the use of symbols in communication in man, whereas such use has never evolved appreciably in even the highest apes? It is commonplace in anthropological discussion to point to the enlargement of the human brain as making speech possible. Generally the basis of the enlargement of the brain is ascribed to other selective forces—tool-use, control of autonomic centers, etc. The selectionist viewpoint espoused here, however, holds that complex structures like complex behavior represent responses to selection pressures specifically and directly favoring them. They do not evolve as incidental accompaniments of other selection pressures. If tool-use, for example, were valuable for foraging animals (which is doubtful, since foraging primates make almost no use of it in food-getting, although quite capable of using sticks, etc., in defense), the kind of mentality it would favor would make for tool-use, not for language (Figs. 10.14 and 10.15).

The concept of an integrated family organization, with economic differentiation of male and female roles as here suggested for proto-

FIG. 10.14. The Galapagos woodpecker finch is using a stick to dislodge insects from a crevice. Though this must be considered a type of "tool-use," presumably it is an inherited rather than culturally determined behavior and is inflexible. (After D. Lack, *Darwin's Finches,* 1947.)

FIG. 10.15. Chimpanzees have a strong tendency (apparently innately determined) to manipulate objects of appropriate sizes by turning them over, stuffing them into available cavities, etc. This is done most effectively in play rather than when the animal is attempting to reach a goal in problem solving experiments. Instances which appear to be "insightful" solutions of problems such as that of fitting two sticks together to reach an object out of reach of either may represent mere coincidences resulting from this manipulative tendency. (After P. Schiller, in C. H. Schiller (ed.), *Instinctive Behavior*, 1957.)

cultural man, points to an important difference from the condition of other primates, which may account for the selection pressure for symbol-formation in man and not in them. In the monkey or ape, each individual forages for himself—except, of course, for infants. For no important period of time are the members of the group out of one another's range in terms of communication. Hence, signals rather than symbols suffice for communication (Etkin, 1963*b*). For example, should an enemy be detected by one monkey, its outcries would serve to arouse the attention of other members who, being in the immediate vicinity, could react appropriately. It is not necessary for the first animal to convey information of a specific kind about objects not present or to specify the direction or nature of the source of alarm. It is sufficient for the animal to express his own alarm and flee in his own way. Thus a signal expressing the sender's own condition, rather than a symbol with a specific referent, suffices. As we have seen in Chapter 3, such signals are widespread among all social animals.

In the integrated family organization the division of labor necessarily keeps the members out of touch. Communication about the hunt—for example, the slaying of a large animal at a distance, requiring co-operation for handling or the injury of a member of the pack—must be transmitted from the hunters to the others. Only symbols conveying specific information about referents not otherwise to be perceived by the recipient will do this. Conversely, events on the domestic scene must be conveyed to the returned hunters. In short, the integrated family with two separate areas of action depends upon the ability to communicate information about things not immediately to be perceived. We may suppose that the special conditions of the integrated-family ecology gave rise to the selection pressure whose outcome was the use of sounds as symbols with specific referents. This communication system arose in one species only because the selection pressure operated upon an animal which was already highly intelligent and which adopted a new mode of living. This unique combination of factors occurred only in the evolution of man.

PLAY

Something has already been said about the special aspects of human play in connection with reproduction. In man as in wolves, the adults, both male and female, play with each other and with the young. Play in man is not only physical but verbal. Social conversation is one of the most important ways of maintaining a friendly social bond and the basis of familiarization (Etkin, 1963b).

Another characteristic of human play is its relation to training for adult activities. The familiar fact that the play of children so often is an imitation of adult activities must be viewed as part of the adaptation of man to cultural transmission. This spontaneous play grades over into definitive instruction and practice of adult roles in which the differentiation of male and female is prominent. As we have previously emphasized, the drives and capacities for learning in particular ways are part of the adaptive mechanisms of an animal. The prominence of the drives to assume and practice adult roles is thus part of man's adaptation to his cultural mode of life. It differs, of course, only in degree from that seen in many mammals. The hunting play of many predaceous mammals, which also includes definitive instruction by the parent, is a case in point. Another is the observation of Fraser Darling that the female yearling in the red deer attend to the reactions of their mothers toward weather signs, whereas the males, never destined to be herd leaders, do not attend and apparently never learn to be proficient.

CULTURAL TRANSMISSION

Cultural transmission, the predominant characteristic of human mentality, has a firm if narrow base in animal behavior. As we have seen, even among birds there are clear examples of the transmission of learned behavior from one member of a group to another (cultural transmission). The learning of foodstuffs and enemies is widespread. One of the most curious examples of such learning is that of the tits and various other British birds which have learned to rob milk bottles by stripping or breaking through the paper tops. This technique appears to have been started by a few birds in an area and spread to the others in the manner typical of the diffusion of cultural elements. As in other cases discussed in Chapter 3, this activity is merely a modification of the normal behavior pattern of these animals, which find part of their food by stripping bark. What has been learned, therefore, is the new stimulus situation releasing the old motor response (Fisher and Hinde, 1949).

In point of fact, all learning involving social relations necessarily assumes somewhat the character of cultural transmission. For example, when a dove is imprinted to its own species and subsequently mates "correctly," it thereby sets up the situation for transmitting this same behavior to its offspring. Whether we choose to call this cultural or "pseudo-cultural" transmission, we must recognize that it has much the same effect as direct teaching in that one generation transmits its learned patterns to the next. Such "pseudo-cultural" transmission appears to play a considerable role in territory, mating, and perhaps other activities in birds.

In mammals cultural transmission of a more conventional character is more commonplace. Since the large social mammals are long-lived, there is much occasion for the direct learning of stimuli regarding territory, migratory paths, food, and enemies. Mammalian action patterns, being less rigidly organized, also are more subject to cultural modification, as we saw in discussing learned hunting techniques. Primates show cultural transmission most extensively. This was particularly evident in the reports from the Japanese Monkey Center [Fig. 10.16]. In apes, though little is known of the details of their behavior in the wild, such transmission of learned behavior is very apparent in laboratory and zoo conditions. Apparently, chimpanzees cannot even mate successfully without instruction or guidance from experienced animals.

If, then, we think of protocultural man under the circumstances (presumably of climatic change) that were forcing him toward a hunting ecology, we may recognize that an alternative to the usual biological method of evolution presented itself to him. Instead of

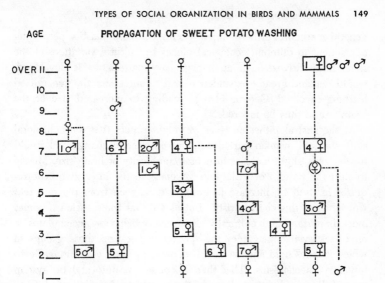

FIG. 10.16. Pattern of Spread of Potato-Washing Habit
in a Monkey Colony

A peculiar habit, that of picking up one of the chunks of sweet potato
and bringing it to some water and rinsing it in the water before eating it,
was started in one of the colonies of Japanese monkeys by a young female
(encircled). This habit was seen to be picked up first by her mother and
two playmates of comparable age. It then spread to other young animals
of the colony, primarily among playmates, but at the time of writing had
not spread to any of the older females or males. The lines indicate gene-
alogical relations from females. The animals taking up the habit are inclosed in boxes and the sequence in which the habit was assumed is given
by the inclosed numbers. (After Miyadi, Proc. XV Int. Cong. Zool., 1958.)

altering each of the behavioral characteristics individually by genetic
change, a good part of the behavioral transition could be achieved by
further development of his already extensive capacity for cultural
transmission of learning. Let us consider dominance behavior as an
example. Instead of altering this behavior mechanism in respect to
the stimuli eliciting it and the motor patterns of its expression by
genetic change of neural mechanisms, the general dominance drive
could be made subject to learned stimuli and methods of expression.
This would lead toward the situation as we see it in man, where both
the objects and the manner of expression of dominance drives vary
with each culture. In this way the alteration of behavioral patterns
by cultural means would become the basis of human evolution. Such
a process of change would be vastly more rapid than the usual

genetic method. Furthermore, it would be more plastic, permitting adjustment to different ecological niches by cultural variations of the same basic drives. Such an evolutionary pattern, as it developed, would demand greatly expanded neural resources for the generalized learning involved. Genetic change leading to increased size of the brain would thus be favored.

At the australopithecine level we might expect that the behavioral shift was still in good part dependent on genetic change and would involve only slight increase in neural complexity or brain size. Merely to shift the pattern of male aggressiveness to the integrated type requires in itself no increase in brain size, as seen from the fact that wolves show this characteristic. But as cultural transmission becomes more important and culturally determined behavior accumulates, a feed-back system magnifies the difference between those groups in which it takes hold and those still dependent upon conventional evolutionary mechanisms. This then becomes an autocatalytic system continually accelerating the selection pressure for further shift to the cultural mode of evolution. It is this dependence upon cultural regulation of all drives that requires increase in the cerebral cortex. Such an autocatalytic system leads to the explosive expansion of the brain that appears to have taken place within a few hundred thousand years early in the Pleistocene period (Eiseley, 1956). The outcome is a modern man who appears to have the same drives as we noted in other animals but little in the way of innate mechanism to recognize the appropriate stimuli or to perform the motor patterns for their expression. What he does have is an enormous predisposition to learn these things from others and to teach them to another generation. Thus the nature of man is to appear to have no fixed nature, only tradition.

Of course, the above remarks will be recognized as highly speculative. They have been included merely to bring out the possibilities of interpretation which consideration of social-behavioral factors add to our resources for comprehending human behavior and evolution. It is this author's opinion that an understanding of human nature must be founded upon an understanding of social behavior in lower animals. Progress in this direction depends, not upon reliance on occasional analogies, but upon recognizing the vast variety of forms that the same fundamental principles can take in the kaleidoscope of nature. It is to be hoped that an appreciation of the ecological principles and the physiological mechanisms in the social behavior of animals will enable us to see more clearly the possibilities for the analysis of our own behavior.

5

A Biological View of Man's Social Behavior

Introduction

We have seen in the preceding chapters the ecological and evolutionary context of social behavior among many vertebrate groups. Man, of course, is no less a product of evolutionary developments than any other creature, and a discussion of man from the viewpoint of the adaptiveness of his behavior can be very revealing.

Unfortunately for the evolutionist man has been so successful an animal that he is the only remaining species in the taxonomical family Hominidae. There is, therefore, no way to reconstruct the phylogenetic history of his behavioral mechanisms via comparisons of living species, as is possible, for example, with ducks (Lorenz, 1958) or lemurs (Jolly, 1967). There are a number of other unique factors in an evolutionary description of man's social behavior; some of these are discussed in the first three sections of this chapter.

Man, Cultural or Biological?

A first objection to the consideration of man's behavior in biological terms is that his behavior is said to be almost entirely culturally determined. Furthermore, it is claimed by some (e.g., White, 1949) that biology may be ignored since culture evolves via its own set of rules. In a similar vein, such well-known biologists as Huxley (1958) and Waddington (1960) have suggested (following Darlington) that cultural transmission is a new form of information transfer; that is, culture is a new step in the evolution of genetical systems. Although there is some truth in these views, they have tended to distract us from the job of extending Darwin's insights (1873) by a careful consideration of the adaptive value of various aspects of human behavior. Thus, we prefer to emphasize that there is no cultural development which is independent of man's biology and that at some level of analysis everything man does has an evolved biological component.

This brings us to problems raised by the cultural diversity of man. That is, one may ask: Does not the very fact of the immense amount of cultural variation (e.g., Sumner, 1906) indicate that in man's behavior biology has a very small role, and that social learning has an overwhelmingly large role? The point is that there is no actual division of biology and social learning, and the apportioning of variance to either heredity or environment is statistical language which only approximates reality; any given societal organization must, in fact, have come about through an interplay of genetic, ecological, sociocultural, and psychological variables in what is best considered as an acausal system of complete interrelatedness—acausal in the sense that no single aspect of the system takes exclusive primacy. Where anthropology has in recent years emphasized the cultural basis for personality, we might as readily emphasize the personality basis for cultural change, and "primary cause" has at some time or other been erroneously inferred at all of the levels of explanation mentioned above, depending on the bias of the investigator —whether he was a biologist, economist, ecologist, sociologist, psychologist, and so on. In reality, the best we can accomplish is to hold constant as many factors as possible in the attempt to examine variation in the others (Barkow, 1967).

Since we are here interested in how the biological level might affect the sociocultural level, let us pursue a hypothetical example. Suppose there are two Polynesian islands, several hundred miles from each other, each populated by groups of five hundred persons, with several islands similarly populated within this several-hundred-mile expanse. On Island A the people are friendly and outgoing; on Island B they are shy and retiring. Let us suppose the founding populations of each island were single families and that friendliness or shyness differentially characterized these two families. Let us suppose further that these characteristics were at least partially based on inherited temperamental differences, an assumption for which there is now considerable evidence (e.g., Freedman and Keller, 1963). Is there any reason to believe that the basic Polynesian institutions will not take on a unique cast as a result of subsequent isolation and inbreeding? On the contrary, it is highly likely that this is part of the process by which a culture gives rise to new cultural forms.

One implication of this reasoning is that the culture developed by an isolated group is particularly adjusted to the gene pool of that group; that is to say, the culture developed by any homogeneous people must reflect the unique biology of that people—as, for example, its temperament. In general, human isolates might in the future best be studied ethnographically, with hypotheses based on population genetics rather than on the relatively shopworn hypotheses of psychosexual fixations. The assumption that all mankind has exactly the same balance of heredity abilities and drives is neither necessary nor probable. An a priori insistence upon such a viewpoint is, in fact, a dogma that can only retard scientific investigation and the fruitful application of subsequent findings to human social problems.

Causality, then, is a relative term in that any level of human organization can affect any other; genes and related temperament may induce behavior which may affect sociocultural institutions which may in turn affect genetic selection, and so on. Because our primary concern is at the level of evolved structure associated with human social behavior, we are focusing upon "biological origins." However, it should be kept in mind that evolution has, in fact, involved interdependence and feedback between all levels.

The Within and the Without of Behavior

As was emphasized in Chapter 3, each animal species has its own system of signals and is especially attuned to attend to them. Man is no exception. We are, as conspecifics, constructed to comprehend each other, and given the proper setting and rapport, we can, for example, quite completely immerse ourselves in the nuances of another person's feelings and thoughts. We cannot do this with other animals to the same degree (although we and dogs have become attuned to each other via genetic selection).

In studying man, if we remain strict behaviorists and deal only with our sense perceptions—as when dealing with inanimate objects or lower animals—we deprive ourselves of insight into vast areas of human experience to which we are naturally attuned, such as affect, imagery, and thought. Thus, in a discussion of the significance of dance and music for a human group, it is a poor ethnography indeed which neglects an explicit appreciation of the joy felt by the participants and the feelings of communion within the group.

In this regard, the animal behaviorist, of course, is not so well off as the student of human behavior. While it is true that ethologists have been interested in the experienced world or Umwelt of their subjects, following von Uexküll (1926), their only recourse has been to attempt its reconstruction via objective experimentation (e.g., Tinbergen, 1953). The shortcomings of this approach become particularly apparent when strict ethological techniques are used on human children, that is, when they are studied only in terms of "objective" perceptions. A trained ethologist, N. Blurton Jones (1966), has made such observations over a three-year period during which he looked at nursery schoolers as if at another species, attempting to describe only what he saw and heard. The net result was an unsatisfactory collection of observations that had little coherence or configuration. It is almost certain that Blurton Jones knew more than he was able to say, given the constraints of the "objectivistic" methodology he used.

One can similarly criticize Darwin's pioneering work, *The Expression of the Emotions in Man and Animals,* for it is clear that he refrained from giving more than a perfunctory description of the subjective aspects of the emotions, preferring instead to describe in detail only the observables. Interestingly, Darwin also refrained from speculating on the adaptive value of man's emotions, perhaps also in the interests of maintaining "objectivity." Given the fact that those were pre-psychoanalytic, pre-phenomenological days, and that the very consideration of man's emotions as species-specific behavior was a major step forward, we can only say of Darwin, as Lorenz has said, that he too "knew more than he was able to say."

While there will be some attempt in this chapter to rectify the scientists' tendency to discuss only the without of things, it will fall far short of the requirements of a phenomenological approach as stated, for example, by Buytendijk (1962). It is hoped that in the future works will appear which combine in a more thorough manner an evolutionary approach with phenomenological analysis.

Innate vs. Acquired

We have been developing a holistic viewpoint by stressing the artificiality of such dichotomies as cultural vs. biological and within vs. without. Monism characterizes nature, and these dichotomies, which are man-made, often mislead us. Another disguise of the cultural vs. biological dichotomy is the perennial opposition of innate and acquired.

The problem with this dichotomy becomes clearer when we consider that the common ethological meaning for innate is "unlearned." Is imprinting, the quick formation of attachments between adult and young precocial birds, innate (unlearned) or acquired (learned)? If this question is pursued at any length, we get into logical difficulties. Imprinting, it will follow, involves the unlearned capacity to learn within a short span of time; that is, temporal quickness is given as structure whereas learning is presumably without structure. There is, however, no logical dividing line separating structure from learning, since learning itself must be a structural product of evolution, and the dichotomy comes to rest at an insoluble impasse. Nevertheless, while we may thus become logically enmeshed with the term innate, in other instances its use is perfectly clear, as with the patellar reflex, for example.

What are we to do about these vocabulary problems? It makes considerable sense, when dealing with complex behavior like imprinting, to speak at a more abstract level, namely, of an *evolved* capacity to imprint. *Evolved* refers only to the fact that imprinting has a clearly perceived adaptive function, but implies nothing about whether or not learning plays a role: it does not suggest opposition with "acquired," as does *innate*, but simply designates probable phylogenetic origin.

Thus, when we refer to evolved or phylogenetically adaptive behavior we are designating a behavioral unit which has been produced by the evolutionary process in much the same way that physical or biochemical characteristics of the species have been produced. We are *not* offering an analysis of the complex interaction of gene and environment that went into the development of the behavior in question any more than we make such analyses in speaking of a physical characteristic such as the pink color of the flamingo. This color is properly discussed as an adaptive characteristic of the species even though experiment has shown it is only expressed under particular nutritive conditions. In the same way, a particular behavior is properly examined from the evolutionary or ecological viewpoint as an evolved unit irrespective of whether the particular genetic and environmental interaction necessary for its expression has or has not been analyzed.

For many ethologists who have become accustomed to opposing innate with acquired, insistence on substituting the more abstract terms "evolved" or "phylogenetically adaptive" is not seen as an improvement. Lorenz, in dealing with recent objections to the use of "innate" by ethologists themselves, has attempted to salvage the term by posing intercalated chains consisting of learned and innate segments (Lorenz, 1965). However, it becomes extremely difficult to imagine such a chain in dealing with, say, the human smile. (See below, "Parent-Infant Attachments.") The fact is that isolation experiments are usually necessary to establish innateness (Lorenz, 1965), but in the highly social primates we know that isolation is entirely antithetical to normal development (Ainsworth, 1962), and the experiment simply cannot be done. At best, we can pose a spectrum of behaviors ranging from reflexes, which are clearly unlearned, through the fixed action patterns of lower animals, in which Lorenz's intercalated chains of innate and acquired segments may be possible, to such phylogenetically adaptive hominid mechanisms as playfulness, smiling, and laughter, which are thoroughly and inextricably enmeshed with learning and experience.

It may be anticipated that a problem with the term "evolved" is that it is frequently used to describe cultural as well as phylogenetic evolution, and the term "phylogenetically adaptive" may therefore be the better of the two. For the purposes of the present chapter, however, these two terms will be used interchangeably.

Dimorphism

Much has been written contrasting men and women, boys and girls, but such differences have rarely been considered within an evolutionary context. The many discussions of the Oedipus complex and how boys and girls differentially master it provide a good example; we know of no such discussion which seriously considers the obvious parallels between human male-male rivalry, which starts among juveniles, and similar behavior seen in other social species (e.g., the rhesus, Harlow and Harlow, 1966).

From the work of Young (1965), Levine (1966), and others we know that the introduction of androgens to young female rats and to embryonic female monkeys permanently virilizes them and, for example, increases their aggressiveness and the number of challenges to fight. It now appears that it is the action of these androgens on the central nervous system during a critical period which produces maleness, and studies of male pseudohermaphrodites have yielded similar conclusions regarding the formation of human sexuality (Landau, 1966).

Thus, it seems likely that the upsurge of rivalrous feelings which male human four- and five-years olds and juvenile primates experience are due in some part to shifts in hormonal levels, most probably androgens, acting upon a virilized central nervous system. The need to win and to be "top dog" seems predicated upon the evolution of social dominance, so widely seen among group-living species, and the infantilized human seems to differ from other primates in this regard largely in the time scale of development. More generally, we cannot persist in the notion that behavioral consequences of hormonal differentiation of the human sexes occur for the first time at puberty, as some writers have held (e.g., Ausubel, 1958).

Although their explanations have ignored evolution, psychologists have gathered a good deal of data on male-female differences. These findings seem always to reflect greater passivity in females and greater aggressivity in males. Even in cultures where women are more active in courting (the Navajo, for example), it is acknowledged within the culture that the males are saving themselves for heroic bursts of energy. There are studies (Ausubel, 1958) which show that boys anger more easily when frustrated either by the activities of another or by the resistance of the inanimate materials they are working on. Young males engage in more rough-and-tumble play, are more highly investigative and intrusive, and seem better able to manipulate mechanically and to visualize three-dimensional relationships (Bock and Vandenberg, 1966). Save for the last point, young primate males and females are differentiated in about the same way.

At about seven years of age human males are more given to forming competitive hierarchies than females; they are more interested in assuming the hero's role, as witnessed by competitive behavior in sports and in their ideation and day-dreaming. Girls seem more oriented toward their adult role as mothers and they seem everywhere to indulge in play which emulates the caretaking role of women. As adolescence approaches, girls participate more and more vicariously in the world of male competition; they seem not to seek the hero's role but are instead adulators of the hero. Studies of adolescent dreams in various cultures bear this out and show also that adolescent boys have more aggressive content in their dreams than do adolescent girls (Booth, 1966). Much of the information on male-female differences is summarized in Maccoby (1966).

It is often claimed that data such as these are not proof of biological differences between boys and girls, since culturally induced parental expectations can lead to differential rearing of the sexes; nor does the fact that similar sex differences characterize many cultures convince such skeptics. It is, therefore, worthwhile to digress momentarily and point out that the way babies are handled is as much determined by the inherited temperament of the baby as by the cultural modes represented by the parents. For example, in studies of infant twins (Freedman, 1965, 1967) it was found that identical twins elicited similar behavior from adults, whether parents or strangers and whether seen singly or as a pair. Fraternal twins, on the other hand, drew consistently dissimilar responses from adults. For example, in one fraternal pair of infant girls one twin became nicknamed Delicata (delicate one) because everyone, including an uncle who had characteristically played roughly with the other children, sensed her fragility and treated her accordingly. Her twin, who was more robust and a striking contrast in many other ways, would laugh only when tossed in the air or severely tickled, and as a result she was treated with greater physical exuberance by all the family.

The point is that among normal parents the temperament and personality of the baby are respected, and infants determine how parents treat them to a much greater extent than is ordinarily recognized. Similarly, with regard to boy-girl differences, current work with opposite-sexed twins indicates that boys, on the average, behave differently from girls from infancy on and that parents respond in terms of these differences.

Continuing our discussion of evolved male-female differences, at puberty the great increase of androgen level in males makes that period particularly touchy in terms of male-male aggression and rivalry. This fact may well be the biological underpinning for puberty rites and male initiation ceremonies, for the rivalrous newly adult males must somehow learn the rules which have been devised to prevent aggression from fragmenting the group. This is discussed in greater detail below, "Group Behaviors."

With regard to pubescent changes in women, menstruation now starts earlier than ever before in history (Tanner, 1961), most probably as a result of improved nutrition and the elimination of many diseases. The flexibility of menstrual age seems to have evolved with the apparent function of limiting or extending the fertile years of females according to the availability of nutriments, which in turn would serve to diminish competition for limited food sources. Many analogous examples in other species, in which reproductive potential is adjusted to ecological conditions, are described by Wynne-Edwards (1962). Human females also mature more quickly than males in all phases of development, such as bone age, teeth eruption, language development, and sexual maturity;

at least one adaptive function of this disparity is that it makes it easier for the males to maintain the just-matured females in a submissive posture since females are younger and less experienced than males at a comparable level of sexual maturity.

While a modern female may bristle at the thought of evolution arranging for her submissiveness, it appears that dynamic dominance-submission relations are the only reasonable means of achieving social stability, given the concurrent adaptational value of a high level of male aggressiveness. On this point, the relatively limited degree of dimorphism among hominids makes it possible for the female to challenge male dominance frequently, a fact of hominid life we usually call the "battle of the sexes." Such challenges are almost always verbal or indirect, and many societies reduce such intersexual conflict by institutionalizing completely separated roles. The psychoanalytic concept of "penis envy" is based on similar data and translates very readily into such a biocultural framework.

Male beardedness and overall hairiness seem best explained in terms of both male-male competition and male-female non-competition. Young men, though strong, are relatively hairless and to some extent the hierarchy based on age is partially stabilized by the tendency to defer to signs of age, including beardedness and hairiness. Awe and respect of the older seem built into man (Waddington, 1960; Piaget, 1932), and beardedness as well as graying may well have evolved as means of designating position in such an arrangement. Among some beardless peoples, such as the Maori, it is interesting that facial tatooing or scarring became a common practice, with the tatoos and scars much more extensive and "frightening" on the males, more decorative on the females. This may in fact give us some insight into the function of beardedness, for the bearded face, too, takes on a forbidding or authoritative appearance more readily than does a non-bearded one (as current research by the author suggests). This is in keeping with what is frequently found in other species—that strong dimorphic traits correlate with the establishment of male-dominance hierarchies (Chapter 4).

It is not directly in line with the present thesis to speculate on the widespread practice of shaving the face, but considering the history of shaving (particularly the Greeks vs. the Romans and the nineteenth vs. the twentieth centuries), this practice seems to relate to cultural emphasis on efficiency, mobility, equality of the sexes, and the inhibition of overt, within-group rivalry.

Lest human male-female differences receive overemphasis, let us again note the moderate physical dimorphism exhibited in Hominidae compared with, say, the seal family (Chapter 4), and the fact that a good deal of exchange of roles, except in terms of warfare and heavy work, is seen in most culture areas. The attempt to obliterate the evolved differences between the human sexes (e.g., Mead, 1939) is, however, by far the more prevalent distortion to be found in recent thought.

Courting

Parental care of the young and maintenance of groups through family ties seem to be the major organizational functions served by permanent pairing (see Chapters 2 and 4). In mankind the probable function of extended pairing is to assure successful rearing of young within a minimal group. Thus wolves need the pack for the hunt, ungulates the herd for protection, and most primates are found in sizable groups probably as protection against predators. Man, given his intelligence, is probably viable at the level of family groups, and prolonged attachments between mating pairs is characteristic. In this section, then, we will consider the means by which males and females first unite.

What are some of the more apparent attributes which attract men and women to one another? In general, men are more readily sexually excited by the visual modality than are women, whereas women first react with comparable intensity when physically contacted. Masters and Johnson, 1965; Kinsey, 1953). A similar dimorphism is found in a majority of species, the male usually taking the active courting role on catching sight or smell of the female. This may be followed by ritualized movements and then by the actual touching of the female's genitalia in a mount. In animals ranging from Drosophila (Spieth, 1952) to the apes (DeVore, 1965), the female's major sexual activity is often restricted to rejection of unwanted suitors.

In each such species, then, there are visual, and often odoriferous, attributes of the female to which the males are attracted. What, then, are the attractants exhibited by the *human* female? The entire concept of female beauty is here our subject matter, and so a complete discussion is not realistic. It is interesting, however, that such a frequently discussed subject has rarely been considered in an evolutionary context.

Regarding the so-called culture-boundness of beauty and disregarding for the moment the obvious fact of gene-pool differences with their associated physical differences, we observe that men the world over seem to have approximately the same taste in female beauty. The marauding armies of history are perhaps the best proof of this. While one often hears that the male Hottentots, for example, prefer women who have grotesquely large buttocks (following Darwin, 1873), the fact is that young Hottentot girls have buttocks which are simply well-rounded and attractive by almost anybody's standard. Only as the women grow older and fat accumulates do they look grotesque to other peoples; but almost all women lose their sexual appeal as they age beyond their reproductive years.

Softer facial features and relative hairlessness of the face and body characterize human females, and Lorenz has spoken of childlike non-angular facial features associated with facial fat, which are retained in postpubescent females and not in males. Similarly, female musculature is softer, giving the entire body a more supple, relatively non-resistant appearance; along with this the relative smallness of females emphasizes their non-competitive subdominance.

The fact that the human female is the only mammal whose breasts are conspicuous whether or not lactating and the fact that breast size and milk production are not related, coupled with the clear attractiveness of breasts to males even in cultures where breasts are not under cloth (Ford and Beach, 1952), is strong evidence that breasts are evolved sexual releasers. That this function of breasts is probably unique to hominids seems directly related to the upright stance, for only in the upright can such a development stand out visually. By contrast many of the quadripedal primates, when in estrus, display a swollen and/or brilliantly colored sex skin as a visual attractant.

In a discussion of breasts as sexual releasers, it behooves us to note that in some advanced cultures, such as the Chinese, large breasts are considered too animal-like, and the practice of binding the chest to reduce breast size has been widespread. Large breasts, of course, are not animal-like, but this practice is in keeping with the general oriental motif of muting emotion and reducing animality and animal drives. Another observation in line with this point is the Japanese disapproval of twin births because the multiple birth resembles the animal litter, and, moreover, genetically selective forces seem to flow with such cultural motifs; this is demonstrated by the fact of reduced breast size among many Oriental groups as compared to other races, and a reduced occurrence of twinning among the Japanese (approximately half the average Negro rate). Along with this last point, certain African tribes which hold twinning in high esteem have a twinning rate about ten times that of the European rate (LeVine, 1967). In the absence of historical documents, it is probably pointless to speculate whether culture dictated genetic selection in these two instances or vice versa—and it is safest to assume a feedback process between the two.

With regard to the phenomenology of courting, the advent of the sweater girl over the entire westernized globe has demonstrated how the breasts can be actively used as an attractant; a lazy stretch, a fully erect back, are all natural responses of seeking females. The consequent male arousal and courting responses need not be documented in any detail, but we should note that once the distance receptors are stimulated and the female is approached, the smile and the meeting of eyes become further media of drawing together. Such prolonged confrontation of the eyes in courtship usually occurs just before intimacy is completed, and while cultural differences enter at all points, the essential nature of the encounter is probably everywhere the same. That the encounter of eyes in the en face position has a good deal of further meaning in human relationships is clear, and it will be discussed again in the section on infant-adult attachments.

It is of interest that female inhibitions in courting are themselves attractive to males, and the girlish giggle and shyness may be viewed as ritualized attractants which indicate a woman's interest in the other sex; the more shy the girl is, as when she looks only partially toward the man, the more she giggles and the more her face turns red, the more attracted he becomes. Thus the reddened face is a message, and during courting the blush and the askance look indicate the female's knowledge that she is being looked at desirously and, at the same time, that she is not undelighted at this attention. The male finds this attractive and, among other things, such behavior never calls into question his dominance, for it is a ready admission of submission. In general, non-cryptic animal coloration is an adaptive attention-getting device, and, psychopathological blushing aside (Feldman, 1962), blushing appears to fulfill such an evolutionary function (see also Darwin's discussion, [1873]).

Inhibitions towards courting seem quite as universal as courting itself. Males, particularly young males, tend to court with an eye on the male dominance hierarchy, so to speak, and feelings of subdominance within the male hierarchy are often associated with sexual inhibitions (see also the last section). To take an extreme but illustrative example, on the Marquesas thwarted adolescent love is often followed by the feeling that peers are derisively laughing at one, and this is the most frequent cause for suicide in adolescent males (Linton, in Kardiner, 1939). The

Marquesas are also illustrative of a society in which adults permit adolescents rather complete sexual freedom and in which adolescent inhibitions, such as they are, are largely self-imposed and based on relative self-esteem within the peer group (Suggs, 1965).

Inhibition to courting is not a uniquely hominid trait, for DeVore (1965) speaks of functional castration (following the Freudian castration complex) in subdominant baboon males who are at once attracted and inhibited about pursuing the female. In humans these inhibitions are played upon differently in different cultures, as are the courtship procedures, but it is probable that they everywhere reflect the male's self-image regarding his position in the dominance hierarchy; so-called adolescent infertility (discussed by Montague [1946]) is probably in some degree due to such functional impotence. It thus does not seem unreasonable to speculate that functional impotence is an adaptive mechanism which tends to assure that the mating males will be the most mature and the most ready to rear a family.

In the same vein, good body build vs. malformation enters into the selective process on the part of either sex, perhaps largely in terms of low self-concept in the malformed leading to non-courtship behavior. For example, it is commonly reported by orthodontists and plastic surgeons that soon after rectification of a jaw or nose, pride of self goes up, and courtship and consummation may follow almost immediately. Dynamic psychiatry has made us aware of neurotically (intra-psychically) caused low self-esteem, but it seems important to recognize the biologically adaptive factor as well.

The Family

In the last section we discussed, however briefly, courtship and the first bloom of love. Contrary to many opinions, love seems to be universally possible, although many cultures consider the choice of a mate too important to be left to adolescent children who find themselves in love. (Goode, 1959). Thus we find a spectrum of possibilities based on the opposing ideals of free choice of mate vs. marriage arrangement. The latter extreme, in fact, does not preclude falling in love, it merely postpones this possibility until after marriage (Davenport, 1965).

Where falling in love is widespread, it seems to occur with greatest emotional pitch soon after puberty (e.g., Mead, 1939) and it is quite clearly a major force through which males and females are maintained as at least semipermanent pairs. This is an important adaptation in light of the necessity for a father-provider in human families due to the prolonged infantilization of the young.

Not all man-woman pairing starts with this high level of emotional feeling, and the attachment process between mates often occurs over a period of time. Certainly an important element in achieving such attachment is the nearly constant (compared to other animals) sexual arousal possible in hominids. In many societies, Sweden for example, the majority of marriages occur following pregnancy, and by that point mates are more often than not deeply attached to one another and marriage normally follows as a matter of course.

It is often pointed out that frequent sexual relations are not necessary for monogamy. For example, in graylag geese sexual activity is perforce seasonal, so that monogamy, if it is to form part of this species' adaptation, must be achieved in a non-sexual way. The way it occurs, according to Lorenz (1966), is via a recurring triumph ceremony in which pairs engage and which clearly involves mutual emotional reinforcement. Primate adaptation, on the other hand, involves the ability to raise young at all times of the year, but the concept that primate groups are held together by sexual bonds (Zuckerman, 1932) is now known to be wrong (DeVore, 1965).

Although temporary attachments may indeed form following coitus in some primates, only when we get to man is there merit in Zuckerman's original notion, primarily in regard to pair formation. The human female has no clear-cut estrus period and she experiences sexual readiness far in excess of other mammals. In addition, as far as we know, only she can experience orgasm—a highly rewarding emotion to say the least. As for males, Kinsey, Pomeroy, and Martin (1948) have reported that coitus may occur ten or more times per night when human males are between fifteen and twenty years of age, tapering off as much as three times weekly at sixty years of age. Thus frequent sexual play and intercourse seems to achieve about the same level of attachment and monogamy in man as does the triumph ceremony in geese.

Beyond the emotional reinforcement provided by sexuality, there are of course many further aspects to partners staying together. Thorough familiarity with someone and time spent together seems to make an important imprint on a human's affective feeling for another. (For a parable illustrating this point, see Saint-Exupéry, 1943). Mutual experience of strong emotions other than those associated with sex often bind people together, so that sharing a frightening, awesome, or joyous experience may reinforce the bond.

In this vein, sexual inhibition and shyness, so far as it creates conflict which is emotionally arousing, probably serves often to solidify attachments among humans; Davenport (1965), for example, has so described the function of proscriptions against premarital liaisons in a Melanesian society. For that matter, sexual inhibition seems to be present in all human societies, and unlike the lower primates (see DeVore, 1965), copulation rarely occurs in front of conspecifics and usually takes place in the dark (Ford and Beach, 1952)—with the exception, of course, of orgiastic ceremonies.

This shyness may well be related to the frequency with which copulation occurs in man and the fact that he is vulnerable to attack, especially by rivalrous males, while he is so engaged. Evidence for this assumption is the frequent fantasy common in various cultures, especially in subdominant males, that another larger male in some way will render him impotent (see, especially, Roheim, 1950). Secretive sexuality serves, as well, to reduce overt sexual rivalry and to that extent maintains the integrated cooperative organization of the family and the wider group. Regarding the possible phylogeny of this behavior, it is important to note that males of many species, particularly those which may be preyed upon, are sexually inhibited when in strange places (Ford and Beach, 1952).

POLYGAMY

The opposition of polygamous drives, especially as seen in males, and the counterpull of developed attachments is another case of opposed drives ending in compromise, since each serves an adaptive function. To spell out what is almost obvious, the polygamous male affords assurance that all females will be fecundated whereas the stable mating pair assures the young of protection and nurture afforded by a father.

For all practical purposes polygamy and polygyny are identical terms, since polyandrous societies are extremely rare (Murdock, 1957), and upon investigation there are always very special circumstances which characterize them, such as too few women (Malinowski, 1931). Needless to say, all societies deal with the opposed forces of polygamy and monogamy in different ways—for example, by legalized polygamy, illegal polygamy, concubinage, enforced monogamy, or prostitution—depending upon the context provided by a given culture. Legal polygamy is present in 418 of the 554 societies rated by Murdock, or in 75 per cent of the world's societies, whereas 25 per cent are characterized by monogamy (Murdock, 1957).

This leads directly into a second aspect of polygamy, that which has to do with male social status, that is, his position in the dominance hierarchy. While this will be discussed later as well, it is important here to point out the fierce possessiveness males have for their women and the simultaneous importance of not being outdone or cuckolded by another male. Thus, being made cuckold is perhaps the most frequent cause for within-group murder over the entire world—and it appears that most cultures exonerate the jealous husband, the initial possessor, who was made "temporarily insane" over his shame and hurt. The deep emotions the human male feels in this situation clearly have as much to do with loss of self-esteem as with loss of the woman's love, and loss of self-esteem is in turn dependent on the male's view of his current position vis-à-vis others in the social hierarchy.

In the same regard, the possession of several women is often involved with inner feelings of well-being—of being "on top of the world"; in many cultures one can witness the importance of male bragging sessions about the number of women one has had, and in some societies the relationship between status and number of women possessed is openly and officially acknowledged. It appears that having the fealty of several women brings out ascendant feelings within a man, as well as the awe and respect of the males around him. Similar feelings and social status may also accompany the monogamous possession of a particularly beautiful or highly worthy female.

It is highly likely that the male's possessive attitude toward the female, once the claim on her has been personally made and/or publicly proclaimed, is an important factor in the various cultural rules concerning incest and exogamy. Thus, if it is not permitted that a clan member come into competition for another's female, much potential bloodshed is avoided, and numerous cultural rules have consequently arisen to prevent such rivalry.

To attribute the universality of incest taboos and exogamy to universal stages in individual development, as psychoanalytic theory demands (the desire of the male child to possess the mother and destroy the father), does not appear reasonable. As stated elsewhere, the fact that two events are related and that one comes earlier than the other does not make the first event causal to the later one. Rivalry between males, and to a lesser extent between females, most likely has its origins in man's phylogenetic history, and it seems to be mediated in part by hormonal effects on the central nervous system. The primary (evolved) emotions of possessiveness and jealousy characteristic of either mate, and a fear of retribution by an outraged or robbed possessor or potential possessor, seems in turn to be the main ingredients which have led to exogamy and incest taboos.

Parent-Infant Attachments

We have so far dealt only with emotions which tie men to women, but the birth of children brings out parental responses in both the human male and female. In contrast to other primate males, who are not outstanding in parental care and are far inferior to the female in this regard, evolution has produced a marked increase in the human male's interest in the young (see Chapter 4). Also, the fact that caretaking is often a common task serves further to enmesh parents one with the other.

In the following sections we will discuss a number of infant behaviors and typical parental responses, and we shall see how parent-infant attachments grow in strength over the first year.

CRYING, HOLDING, AND CARETAKING

The very first behavior exhibited by the human newborn is the cry. This is a common mammalian occurrence and seems to serve the general mammalian function of exciting the parent to caretaking activities. In dogs, for example, a puppy removed from the nest immediately starts to cry and continues until exhausted. The bitch will usually become extremely excited, seek the source of the cry until the puppy is found, and then fetch it back. What we have here, clearly, are two complementary evolved mechanisms, neither of which has to be learned.

In the human species, similarly, it can be demonstrated that within hours after birth most crying infants will quiet when held and carried. Consider how this cessation of crying coordinates beautifully with the intense anxiety felt by the parent until the infant is quieted. Aside from caretaking and feeding, body contact is the inevitable result of crying, and the human baby does as well as the macaque in getting next to the parent even though it lacks the ability to cling. There seems little doubt that such contact is normally a mutually reinforcing experience, and tactile contacts of one form or another remain an important means of relating throughout the lifespan.

SMILING

Smiling is also quite clearly an evolved mechanism (Ambrose, 1960; Freedman, 1964). It is universally present in man and it has the same or similar interpersonal function everywhere, that of a positive greeting or of appeasement. Smiling is first seen in reflexive form in newborns, including prematures, when they are dozing with eyes closed, usually after a feeding. Even at these early ages, however, smiles can also be elicited by a voice or by rocking the infant. Since it occurs in infants whose gestational age is as low as seven months (Freedman *et al.*, 1966), it would not be surprising if smiling, like thumb-sucking, is eventually found to occur in utero. Visually elicited smiles occur somewhat later than aurally elicited ones, though they are occasionally seen within the first week of life. These are called social smiles, since they occur most readily when the eyes of infant and adult meet, but in the auditory mode too the preference for a voice over other sounds also marks such smiles as "social" (Wolff, 1963).

The major function of smiling, then, from a very early age is responsivity to another. It provides an important means of attachment between adult and infant, and in later life it lends ease and promotes attachment in a wide variety of social encounters. It is also widely displayed between adults as a gesture of greeting and appeasement, and it is a major means of either precluding or overcoming dissension and angry feeling.

It is pure surmise, of course, whether the smiling response appeared phylogenetically as an adult-adult mechanism or as an adult-infant mechanism. It is most akin to the "frightened grin" in other primates, a gesture frequently made by a subordinate animal when passing close to a dominant one (Hall and DeVore, 1965), and human smiling may well have originated with such a gesture in an evolutionary "turning to the opposite."

WATCHING, COOING, LAUGHTER, AND PLAY

The importance of the auditory and visual receptors in the young human infant seems directly related to its general motor immaturity. Thus the eyes begin to search for form and movement in the environment soon after birth (Fantz, 1963; Greenman, 1963), and by two weeks of age over 50 per cent of all infants will visually follow a moving person (Bayley, 1961).

At about two months the infant's searching for the adult face can be very impressive. If held at the shoulder an infant may hold its unsteady head back to get a view of the holder's face, craning its neck like an inquisitive goose. One is left with the ineluctable feeling that searching out the en face position is itself an evolved mechanism. Supporting this contention are several experimental studies which find the face a preferred stimulus for most infants, including newborns (e.g., Fantz, 1966), and the fact that the adult feels "looked at" for the first time just before the onset of social smiling. The human orientation toward the face of another is undoubtedly bound up with many aspects of evolutionary adaptation, including the upright stance, the relative hairlessness and rich musculature of the face, and, perhaps most importantly, the high degree of interpersonal communication.

A few weeks after en face smiling starts, the infant begins to coo at the beholding adult who, in turn, usually feels an irresistible urge to respond, and as a result much time may be spent in such happy "conversation." Feedings and sleep have by then decreased, and normally more and more time is spent in direct social interactions.

A more robust order of interaction is initiated by laughter, usually at four months, when the baby and caretaker begin to engage in mutual play. The joy the adult feels in this engagement is probably no less an evolved mechanism than the laughter of the baby, and doubtless such mutually reinforcing emotion results in attachment.

The factor of time spent together is also a solidifier of attachment and this is served, of course, by all the mechanisms, described above.

FEAR OF STRANGERS

As the infant becomes embedded in the lives of those about him, another common phenomenon emerges—the fear of strangers. As early as three months of age in some infants, a definite preference for a parent or caretaker may be seen. This may be manifested at first by preferential smiling and cooing and following with the eyes. The infant may then cry when confronted by a stranger, especially if the place is also novel, as in a doctor's office.

The possible phylogenetic origins of this response are suggested by the fact that many mammals and birds show similar fear responses to strangers and strange places after they have formed their initial attachments. In carnivores, for example, the fear response starts as they begin to travel farther and farther from the nest (about five weeks of age in dogs). Closely related is the advent of fear of heights, which follows soon after the beginnings of motility in animals and humans, and without prior experience of falling (Gibson and Walk, 1960). In general, all animals become exposed to many new dangers as their investigative drives take them from the nest, and such self-protective counterdrives assure survival.

Although motility and fear of strangers are related in lower mammals, a human infant usually develops its fear of strangers between six and nine months, when it does not have the motor ability to escape a predator. Therefore it is a reasonable hypothesis that in human infants the fear of strangers serves mainly to prevent dilution of primary relationships and, in addition, serves to intensify the bonds between the infant and those already close to him. In this regard the experimental work of Kovach and Hess (1963) with chicks suggests that the reaction of fright can make early social bonds even stronger, so that a similar function is already served in lower forms.

IMITATION AND LANGUAGE

Toward the end of the first year of life, human infants start imitating the now-familiar adults and children around them. On the other hand, infants who have had no opportunity to form attachments become withdrawn, apathetic, and uninterested in those about them (Ainsworth, 1962), as do similarly deprived rhesus monkeys (Harlow and Zimmermann, 1959; see also Chapter 4), so that attachments are a necessary prelude to autonomy, investigativeness, and social imitation.

Imitation is clearly a magnificent means for rapidly acquiring rudimentary skills. Complex motor activity, facial expressions, and language are largely dependent on this ability to imitate, and these acquisitions in turn provide the structure within which self-propelling, autonomous activity eventually becomes dominant. Thus the ability to imitate is clearly a mechanism with great evolutionary (adaptive) significance.

There must, of course, be a pre-potency or pre-programing for all things imitated, as, for example, the use of language. One aspect of the pre-programing of language is the fact that the human mind has a time- and space-binding quality, which we may conceive as having arisen under the selection pressure for complex communication about events not immediately present. Thus, what differentiates human language from the communication systems of animals is the fact that the signals of true language refer to and evoke concepts of things not present in the senses of the communicators, whereas in most cases animal signals simply direct attention to things present (Etkin, 1967). More precise remarks are probably not possible at this time, but Chomsky (1965), for one, is currently attempting to spell out how pre-programing of langauge is transformed into usage.

AUTONOMY AND ITS COUNTERFORCES

In those children who have formed attachments, the drive for independent action becomes insistent as motor independence is achieved early in the second year, and foolhardy bravery, extreme negativism, and possessiveness appear almost simultaneously (Ausubel, 1958). As a consequence parental watchfulness becomes extremely important at this age, and paternal protectiveness may now be more frequently seen. It is interesting that in monkeys, such as the Japanese macaque (personal observations), the dominant male develops a ferocious protective response to the newly motile young; and, in fact, since these monkeys travel on the ground as a troop, it becomes largely his job to protect against possible predation. Similarly, while protective responses on the part of human adults may be experienced as purely volitional acts, the intense emotions and vigorous activity aroused when a child is endangered seem, instead, reflexive in nature.

Besides the protection provided by a parent, toddlers have numerous phylogenetically adaptive responses which offset the dangers contingent on investigativeness and motoric bravery. As discussed previously, the fear of falling from heights is quite clearly "built into" most animals, including the human (Gibson and Walk, 1960); further, no matter how brave the toddler, he usually melts into loud sobs when he finally realizes he is lost. Also, various fears become characteristic of most children; fear of large animals and kidnappers, fear of being lost, nightmares of being captured and eaten, appear not to have exclusive origin in parental warnings, although they may become exacerbated by them (Jersild, 1954). Thus, constant readiness for danger seems to typify man even as it does all animals, particularly those which may be preyed upon.

An additional counterforce to egoistic expression is the tendency toward obedience and acceptance of wishes and commands of parents, and, to a lesser extent, of older persons. Waddington (1960) has even proposed that there are biological roots to the "ethicizing" nature of man, and he contends that obedience, morality, and ethical behavior are in man species-specific behaviors. In support of this position are Kohlberg's (1963) findings that moral judgment develops in an invariant sequence within the various cultures and subcultures he has studied. Although Kohlberg does not interpret his results in this way, Guttman (1965) contends that such cross- culturally invariant sequences are evidence for biologically rooted phenomena.

PEER GROUP FORMATION

As already discussed, rivalrousness with and orientation toward peers is seen in many juvenile primate groups (e.g. Harlow and Harlow, 1966), and by four years of age it may be clearly seen in human male children. As in many other primate groups, female rivalrousness is relatively dilute and is most often directed towards other females.

Male-male rivalrousness, as schooling age appears, may take many forms—achievement, proficiency at games, and so on. What typifies much behavior at this time, say five years to puberty, is the increasing importance of the peer group as judge and source of a child's feeling of well-being. If he does well in their eyes, that is, if he rates high in the hierarchy (of which one need not be clearly aware), his self-esteem is boosted, and the opposite is just as true. While parents continue to exert an influence to a greater or lesser extent, depending on the culture, the peer group takes on great importance in human life during these years.

Group Behaviors

It should present no logical problem for evolutionary thinking to consider groups of animals, either conspecifics or in the case of commensalism non-conspecifics, as having evolved mutalities and group configurations which in turn give each individual greater viability. Darwin's (1872) discussion of the probable evolution of neuter forms within various species of Hymenoptera is a classic discussion of this point.

With these facts before me, I believe that natural selection, by acting on the fertile ants or parents, could form a species which should regularly produce neuters, all of large size with one form of jaw, or all of small size with widely different jaws; or lastly, and this is the greatest difficulty, one set of workers of one size and structure, and simultaneously another set of workers of a different size and structure;—a graduated series having first been formed, as in the case of the driver ant, and then the extreme forms having been produced in greater and greater numbers, through the survival of the parents which generated them, until none with an intermediate structure were produced (p. 272).

In this way, Darwin concludes, we can account for the evolution of neuter insects as individuals who lend the species greater viability but who cannot themselves reproduce. The logic is not dissimilar in considering the origins of mechanisms or traits which have maintained viable mammalian groups. The birth, say, of altruistic individuals might give the entire group greater viability, thus preserving the gene pool, even if the initial altruistic individual did not survive. The presence of aged persons within the group may be similarly considered. Although themselves beyond their highly reproductive years, their wisdom, experience, and leadership have probably given human groups considerable adaptive advantage, so that the gene pool containing genes for longevity has tended to persist. Additionally, since it is necessary to care for human infants for a substantial period, longevity beyond the fertile years, particularly in women, would allow rearing of the final child. Thus, although present-day nutrition and medicine have certainly increased the proportion of old people, fossil finds indicate that longevity has long characterized hominids; furthermore, it appears that females have always outlived males in the human as well as in many other species (Birren, 1964).

Particularly at the outset of this section on group processes, it seems necessary to present a disclaimer: I shall be dealing with subjects which have always concerned learned men and a full discussion is here out of the question. Given the enormous complexity of the topic and my relative unfamiliarity with the literature on group behavior, this section is included because of a conviction that the evolutionary point of view can help clarify a number of important issues.

It also seems worthwhile to point out again that none of the concepts under discussion are independent of each other: dominance-submission, amity-enmity, obedience-autonomy, aggression-appeasement, love, parental responses, possessiveness, territoriality, and so on. All are coexisting part characteristics of man, and are convenient nominatives, necessarily used one at a time. Various publications have sought to show that man is basically or primarily one of these, for example, territorial (Ardrey 1965), with the rest considered as background. While it is true that life is whole and writing can never be, the evolutionary approach can only be weakened unless the attempt is made to interconnect trends and to picture at least the filmy outlines of the whole.

DOMINANCE-SUBMISSION HIERARCHY

All primates show various forms of social interactions and there can be little doubt that man's social interactions are also based on related evolved mechanisms. One outstanding similarity among almost all highly socialized vertebrates involves the establishment of a dynamically stable dominance hierarchy. Pertinent animal data have already been discussed at sufficient length in preceding chapters in this book so that it remains to describe hominid hierarchies.

Ascendancy and submission are everywhere present in the institutions, concepts, and activities of man. Osgood (1963), for example, has demonstrated that an extremely high percentage of the emotional words of many languages in many culture areas involves references to relationships of power—ascendancy over others or submission to others.

We have already discussed the rivalrousness of the four-year-old male and its probable linkage to a rise in androgens and suggested that such rivalrousness comes to be expressed in competitive games, usually with peers. At first the rules are unimportant, and almost all five-year-olds are primarily interested in winning, often becoming tearful and hurt at losing when the winner is older and larger, and angry when the winner is younger. (See especially Piaget [1948] for an account of how adherence to rules gradually develops.) In the same vein, there is fairly constant self-evaluation vis-à-vis peers and others with regard to strength, height, agility, and mental capacity. For example, who can urinate the farthest or who can jump the farthest are common boys' games in many cultures. Needless to say, different cultures and different families attempt either to modulate or to encourage competitive behavior, but no culture has been described in which male-male competition is specified as absent.

Another aspect of belonging to the group is a member's pleasure in the group's achievements and status vis-à-vis other comparable groups. This principle appears to hold between individuals, between families, between groups, and between clans. All want to be first and to be part of something which is first. We might add that the Marxist ascription of such behavior exclusively to competitive capitalist society must be modified since in some form competitiveness probably typifies all societies.

As mentioned previously, females often achieve status vicariously through their mate and their children, and they do not necessarily experience this as second-best achievement; it seems, by and large, more characteristic of them not to vie directly for a position in the social hierarchy. Instead, as in baboon groups (DeVore, 1965), the female vies with other females in terms of position of the males associated with her (for example, husband and son) in the dominance hierarchy.

Many familiar behaviors appear to stem from the hierarchical arrangement of groups. Looking at others and being looked at are very important in hominid life and seem to reflect the importance of assessing oneself in relation to others in the group. It is likely that clothing, adornment, self-grooming—indeed, the evolution of "good looks"—are associated with this aspect of hominid adaptation.

Along with the constant looking at others and assessment of their externals, humans have a correlated interest in what others are thinking, that is, the assessment of their thoughts. With this tendency goes a sensitivity to public derision, familiarly experienced as the tendency to see derision in what may be the innocent laughter of a group of strangers. To call this process the projection of one's own derisive feelings, as parlor psychoanalysts do, neglects the non-pathological, socially adaptive aspect of such behavior. Clearly, to be socially shamed is a powerful deterrent to group-disrupting activity.

Finally, it should be emphasized that wherever the dominance-submission hierarchy is found it serves to prevent excessive conflict in that it offers a way of stabilizing a group (chapter 1). The moments of active challenge are highly dramatic and therefore often receive emphasis, but by and large viable groups are those which achieve substantial stability and cooperation within a mutually accepted hierarchy, where divisive tendencies are balanced by integrative ones.

ONE-UP VS. ONE-DOWN

The subjective aspect of interpersonal encounters among hominids often involves feelings of being one-up or one-down, and these feelings seem to stem from the hierarchically oriented nature of the species. To give a common example, in either purchasing or selling, avoiding the one-down position and attaining the one-up position is characteristically more important than the amount of money gained or lost; the proof is that often the very same subjective feelings (ascendancy vs. hurt and anger) follow either petty or substantial gains or losses. Psychological sub-theories are often based on this fact, and Festinger's (1966) "dissonance" theory is based on the observation that persons attempt to justify acts already completed, that is, they attempt to make sure they are one-up or else attempt to reverse one-down situations. For example, one often window-shops *after* making a purchase for reassurance that it was a "good buy." Berne's (1964) descriptions of "games" people play with each other also tend to be examples of persons vying for a one-up position. In the same vein, Haley (1960) has graphically described the psychotherapeutic treatment of a neurotic as the vying between therapist and patient for the one-up position. Haley wryly concludes that when the patient finally realizes he cannot become one-up on the therapist, because of the very nature of the doctor-patient relationship, it is called a "cure."

The same emotions seem to characterize negotiations between groups. In collective bargaining between unions and employers, or in tariff discussions between nations, and so on, it is of considerable importance that each side leave the bargaining table with the feeling that it has gained something; if there is the feeling by one party that it has not gained, or gained less than the other party, hostility usually again erupts within a short period of time (Sawyer and Guetzkow, 1965).

Thus the sense of being one-down and the consequent attempt at realigning the position of power (one-upmanship) is a driving force in many hominid activities. So far as I can see, there are no facts which support the notion that this competitive relationship with other males and other groups ever dissolves into pure amity; and, in fact, amity without enmity is found in no primate group (DeVore, 1965; Ardrey, 1966).

FRIENDSHIP

As we have seen, major drives are almost invariably modified by other equally significant drives, and friendship and affection between peers are the most frequent counterforce to competitiveness and enmity, with both sets of emotions frequently existing side by side. Developmentally, friendships among peers form during the same period (five through ten years) in which rivalry with peers is established; these relationships slowly replace the primacy of parental ties (even as they do among rhesus monkeys), and many friendships formed at this stage last a lifetime.

One can reasonably speculate on a number of phylogenetically adaptive functions served by the tendency to form friendships. Having a close friend of his own age facilitates the child's exploration of the environment and helps overcome his fear of the unknown. Also, such a relationship can provide security; the child can relax his vigilance and not take competitive encounters as seriously since his self-esteem is not constantly in question. In the same way, he can learn from a friend more readily because he is not so fearful of exposing his weakness or ignorance; similarly, both friends can feel freer to mutually explore the world of inner experience. This latter point is particularly evident in the many hours pre-adolescents and adolescents spend discussing their emerging sexuality and sexual interests, and in the mutual support and exploration that characterizes friendship at these ages.

The depth of feeling and communion which may envelop two persons has been described by many (see especially Buber, 1958), and it seems likely that human communication is largely based on this capacity to reach out in a positive way to another, whether friend *or* stranger. For new ideas may be explored only when superimposed on such an emotional base of positive feeling; otherwise there is only information transfer.

In the words of Buber:

"As experience, the world belongs to the primary word *I-It*. The primary word *I-Thou* establishes the world of relation (p. 6).

It is simply not the case that the child first perceives an object, then, as it were, puts himself in relation with it. But the effort to establish relation comes first—the hand of the child arched out so that what is over against him may nestle under it; second is the actual relation, a saying of *Thou* without words, in the state preceding the word-form; the thing, like the *I*, is produced late, arising after the original experiences have been split asunder and the connected partners separated. In the beginning is relation—as category of being, readiness, grasping form, mould for the soul; it is the *a priori* of relation, *the inborn* Thou (p. 27)."

LEADERSHIP

The pull toward the group and toward others is almost always countered by drives toward independence or leadership, and the dynamic combination of such apparently opposed drives gives the species an essential part of its viability.

Leadership of a group is probably an open or secret aspiration of all hominids, and there seems to be a sizable psychological gap between the alpha and beta positions; beta or lower positions seem always to be experienced as a compromise between desire and possibility. Perhaps largely as a consequence of this great valuation, the leader is held in awe—and the greater or more powerful the group he leads, the greater is the awe in which he is held. That there is an additional "pre-programed" aspect to the awesomeness of the leader seems reflected in the child's implicit respect for the stronger, wiser, and older individual (see especially Waddington, 1960). Culture seems to have fostered this built-in tendency with ritual and with the various trappings which surround the leader: the wealth, ornate dress, and attempted deification of royalty are well-known examples.

As to the question of age and leadership qualities, recent work by Neugarten and her associates (1966) indicates a progressive introversion and reduced sensitivity to actual or perceived group pressures as humans of various cultures enter their middle years. The development of such inner-directedness may well enhance leadership qualities, but whether or not this trait has been arranged at the biological level through adaptive selection deserves more discussion than I can here give it; it certainly appears so.

Finally, it should be emphasized that although the attainment of leadership is a motive which may become more intense the closer the individual gets to the alpha position, the most important thing is that the individual "find" himself within the group; any place in the hierarchy is experienced as better than none, and the omega individual strives to stay within the group as much as any other individual (see Chapter 2). This fact is dramatically revealed in the extreme social conformity experimentally obtained by Asch (1952) and others; in these studies naive individuals, rather than risk ostracism, accept the obvious distortions of reality held to by a rigged group.

SEXUALITY AND GROUP COHESIVENESS

Extrafamilial groups normally form along kinship lines, either patrilineal, matrilineal, or both; or else arbitrarily, as in the Australian totemic system. The awareness of one's position in a group wider than the family usually takes place at puberty, and many cultures feature puberty rites to accentuate this. However, even in the absence of such rites, a gradual immersion into a defined group occurs during adolescence (in distinction to the informal groupings of preadolescence). In San Francisco, for example, I have observed Negro and Caucasian preadolescents congregating freely in the same groups. With the advent of sexuality, however, this is only rarely seen; it is as if in-group vs. out-group has become defined along color lines. The badge of color seems here to be no more than a convenience whereby these boundaries between groups are established, for groups can just as easily form along such dimensions as language (including dialect), religion, ancestry, political party, or economic class. The point is, however, that same-sex groups form as interpersonal aggression rises in males consequent to the hormonal changes of adolescence; aggression may then be turned toward outsiders even as reasonable in-group amity is retained.

As discussed previously, the all-male group becomes so important in adolescence that the search for females takes place with a constant eye back toward the male group. Status and rivalry thus enter the courtship and sometimes displace in significance the sex object itself, as for example when young men, of Western and non-Western civilizations, engage in all-male bragging sessions about women they have had, or when they engage in such sexualized competition as comparison of penis size (see Suggs, 1966, for a non-Western example).

Freud (1922) was perhaps the first to point out the relationships between sexuality and group cohesiveness, and the fact that male solidarity largely depends on holding in abeyance aggressive competitiveness. Also in keeping with this view is the fact that male homosexual activity is not uncommon during this period in which male competition and sexuality can become confusedly intertwined; although its etiology is still largely unknown, adult male homosexuality seems to involve in part the persistence into later life of this confusion of sex and rivalrousness, and fellatio, a common homosexual practice (although a frequent heterosexual one as well), is probably always accompanied by intense feelings of dominance or submission.

TERRITORIALITY

Ardrey (1966) has pointed out that a nation attacked by an objectively stronger nation may fight with unpredictable ferocity, and that this is the rule rather than the exception. There is, in fact, no doubt that love of country is in the human a deep emotion that turns to intense anger when the country is under attack. Ardrey relates this phenomenon to the observation that animals defending their own territory are considerably more ferocious than animals intruding into the territory of conspecifics.

Territoriality in its "purest" hominid form seems to occur in hunting peoples. The Ute Indians, for example, had a fairly well-defined territory in the Southwest of the United States; they were ferocious toward intruding hunters from other tribes and only slightly more tolerant of the non-marauding Pueblo tribes. Vayda (1961) has described much the same phenomena among the Maori, emphasizing the aggrandizement of food-producing land as the major cause of warfare among these Polynesian peoples. Such a situation seems to warrant the classic evolutionary explanation for animal dispersion: territorial aggrandizement results in species dispersal over a considerable area. In this way the weakest members or groups are relegated to the least desirable territories, and a balance between food supply and population growth is thereby maintained (see Chapter 1). The higher primates, however, are not in fact highly territorial (DeVore, 1965)—in contrast, say, to songbirds or rutting herd animals, in which such an arrangement is quite clear-cut; in man territoriality, possessiveness, and social status are inextricably bound up with one another.

Teen-age gangs in United States cities are a case in point: typically there is a territory with definite boundaries within which only members of the resident gang can move with impunity. All youngsters of similar age in a strange territory must perform certain rituals to avoid attack. They must not travel in packs, must avert their eyes from residential gang members, and in general assume a submissive mien. Although this does not guarantee safety, it helps. In a neighboring territory, of course, the roles are reversed.

Written signs with courageous slogans serve as advertising and demarcations of territory as well as challenges to rival groups, very much as scent-marking is used, for example, by the lemur. In addition to holding a territory, however, gangs must demonstrate their superiority over rival groups and this sometimes involves either prearranged or spontaneous battles; in this way a hierarchy among groups is in time determined in various sections of the city on the basis of relative courage and size of the group, even as Koford (1963) has noted as occurring among the rhesus of Cayo Santiago Island. Also, as in other primates, an adolescent female tends to derive status from the position her male or her male's group holds on the hierarchical ladder. Organization within a territory can be very strict, and punishment is usually meted out to non-participants, so that the external pressure for an individual to join a gang is very great and the territory is thereby kept free of non-participants.

This is far from being strictly an adolescent phenomenon, of course. In universities and within large businesses the acquisition of space often becomes a major factor of dissension among members of the faculty or business hierarchy, since the size of one's office or "empire" is taken as a measure of one's hierarchical position. Again, this exemplifies the unclear delineation between territoriality and status in hominid adaptation.

Finally, it may be that in the human, as in other animals, a sense of spacing—that is, a sense of minimum space in which one feels he can adequately engage in his daily activities—is an evolved mechanism. But a careful analysis of how homeostasis (and therefore viability) might be aided by such a mechanism has yet to be made.

EXTERNAL THREAT AND GROUP COALESCENCE

As mentioned above, the usual reaction of group-living species is for internal cohesiveness to go up as external threat rises (e.g., Ardrey, 1965). The extreme sensitivity of this highly adaptive mechanism is borne out by the fact that the experience of threat need not necessarily coincide with reality. For example, the peaceful marching by Negroes in white areas in Chicago in the fall of 1966 in the cause of open housing met with the most violent of emotions—hatred, anger, and murderous threat. Whites who ordinarily went their own ways unified under the perceived threat of a Negro invasion. The point is that this intense vilification of the Negro marchers occurred without much possibility that a Negro would actually move into those areas; it was the very thought that led to the cohesive outburst against the imagined forces of evil and decay. As mentioned previously, in-groups may form along different lines, and education can influence how they became constituted; it is nevertheless clear that in-groups always form, and it therefore appears to be part of man's nature.

With regard to ethnocentrism or in-group allegiances, Ardrey (1966) and LeVine (1966) have pointed out that groups with considerable internal divisiveness form into armies only with difficulty, whereas groups which suppress internal dissent tend to form active, marauding armies. Although it is true that any group under attack, whether it is divisive or cohesive, will close ranks to meet the danger, only internally cohesive groups tend to seek out wars. It is as if warfare, or aggression as a group, is necessary to maintain the internal intactness of such groups, and Schumpeter's classic essay on imperialism (1951) is, in fact, a documentation of this point.

Thus, although aggression and rivalrousness may be differentially channeled, depending on the way a culture has arranged for group identifications, the overriding point is that aggression and rivalrousness seem to characterize all groups, albeit the size and composition of the in-group may vary. It has been said before that peace on earth will occur only after an interplanetary rivalry has been established, and to some extent that is the present conclusion.

DISPLACED AND RITUALIZED AGGRESSION

Much of man's life involves linguistic contacts, and many activities which appear as motor acts in lower species appear linguistically in man. For example, language usually plays a substantial role in aggression, court-ship, and greeting behavior, and the specific vocabulary is incidental to the motivation underlying these behaviors. Similarly, *displaced aggression* is almost always verbal in man, and we usually call such behavior gossip, grumbling, or scandal.

To give but one familiar example, in an organized hierarchy, such as an army or a police force, an official system of deference that gives naturally occurring enmities the overt appearance of smooth operation is in force, and within-group aggression must be displaced into gossip, grumbling, and scandal.

Regarding the general importance of these behaviors, Gluckman (1963) has pointed out their great prevalence among primitive as well as advanced peoples, noting that they serve to maintain overall social equilibrium and, ironically, social unity. It is clear that everyman enjoys the verbal demise of the great, and the more vicious the gossip and regal the person, the greater is everyman's pleasure. Again, this is one side of our nature, for we also decry blatant forms of scandal as being in poor taste, and the successful gossip must hew the narrow area between.

Gluckman has also pointed out that one of the most important parts of gaining membership in any group—for example, in becoming a pro-fessional in an academic field—is to learn its gossip and scandals. Fur-thermore, it is an unwritten rule that gossip and scandal are shared only among insiders, and the stranger who engages in them usually commits a faux pas; for in hominids the need to defend one's group from *external* depreciation is intensely felt.

Witchcraft, voodoo, condemnation to hell, and related notions, are cross-culturally recurring forms of *ritualized aggression*, and they usually have highly symbolic, religious connotations. Their social function is not unlike scandal in that, when taken seriously by a society, they tend to make everyone behave in socially acceptable ways.

Sports are quite another order of ritualized aggression; normal com-petitive feelings are formalized via rules and regulations so that danger of mayhem is reduced, and in this way they provide a relatively safe forum for competitive expression. Chess, for example, is one of the world's most widely played one-to-one games, and in psychological studies it has been found that players often experience a fierce desire to destroy or render impotent their "regal" opponent, all within the relative gentility and intellectuality of the chess board (Fine, 1956). As for team sports, although there is a certain pleasure in seeing a team work to-gether, the major affective experience of both audience and players seems to be the stimulation and ascendancy experienced in the defeat of opponents. It is well known that aggressive feeling can run high among spectators—to the point that mayhem may occur, for there are no referees or specified rules in the stands. In general, it seems that no human activity other than warfare receives such massive interest and allegiance in modern societies as do team sports, and hero worship here reaches its most fervent heights. Only on the sports field do men come so close to mayhem and yet usually avoid it, and it seems that basic hominid motivations are involved.

It is of interest that the forms which team sports take seem closely related to their probable inception as ritualized competitions among hunters and warriors. Most team sports emphasize running and accurate propulsion of missiles, and these games seem to have derived from every boy's pleasure: running, climbing, and the accurate throwing and dodging of stones, sticks, and snowballs. It is pertinent that accurate throwing is also found in nature among other primate species (Kortlandt and Kooij, 1963), and it is probable that hominid youngsters, like chimpanzees, take readily to stick and stone weaponry as a result of hominid evolution (Dart, 1956).

Bullfighting is a unique sport in that bravery and courage and derision of cowardice are the major emotions experienced, and the hero image, although it emerges in all sports, is here directly sought so that it may be vicariously experienced by the spectators. As we have already discussed, here too is a basic hominid characteristic found predominantly in males: the need to be esteemed by the group for one's bravery and heroism, as well as the need to esteem another for those same traits.

Finally, gambling is a widespread form of direct competition which is practiced in both primitive and industrialized societies. It is a rather safe, yet highly stimulating, form of competitive encounter and seems to serve the same general needs and functions as sports: one wins or loses and experiences ascendancy or depression without disruption of the ongoing social structure.

GREETING AND APPEASEMENT BEHAVIOR

As we have seen, hominid aggressiveness is a major disruptive factor and one which, unless properly channeled, leads to group disintegration. The history of unsuccessful communal colonies in the United States is also, in large part, a history of the disruptive forces of interpersonal anger (Wilson, 1955). In a recent relatively successful religious colony, the Brüderhof, a major reason for continued success has been attributed by the participants themselves to a single rule: all feelings of anger were openly and immediately dealt with, for only in this way were its fragmenting effects avoided (personal communication).

In most cultures and for most groups such a rule is unworkable, however, owing to well-established deference systems and deep inhibitions against an admission of anger toward those higher in the social hierarchy. Thus, in addition to gossip and other forms of displaced aggression, most human groups depend on greeting and appeasement ceremonies to maintain social bonds even as analogous ceremonies are necessary for viable social functioning in other species (Chapter 3; see also Lorenz, 1966).

Goodall (1965), for example, has observed greeting and appeasement behavior among free-living chimpanzees and has been impressed with their "human" qualities. She has described the tension within a group as a dominant male enters it—until he shows to the group he is not in an aggressive mood. He may show this by touching others on the body or hands, or by nuzzling, whereupon the group visibly relaxes.

Greeting ceremonies in man are, needless to say, quite similar. The touching or raising of open hands is a frequent form of greeting, as is the kiss, an embrace, and submissive lowering of the head. The display of the smile can be perceived over a considerable distance and is therefore particularly effective in reducing tension. Relaxed posture, of course, is also important, and a stalking posture is readily perceived. Words of greeting are equally important, and a threat vs. a friendly greeting are usually distinguishable even if the language is not known.

Like other animals, man can be deceptive, and his greeting ceremonies are sometimes used for deception. Most often, however, they become ritualized as an everyday lubricant in social relationships. Consider, for example, a public argument between two males, or an athletic match such as tennis, and the importance attached to the two contestants' exchanging a few friendly words afterwards to overcome the engendered tension. If mutual appeasement does not occur following an aggressive encounter, tension continues and anger builds. It is clear that in man a smile, a touch of hands, or a few ritualized words can do wonders in dissipating growing anger.

Many cultures institutionalize smiling, touching of hands, kissing, embracing or bowing as parting-greeting ceremonies, as in France or Japan. Significantly, relationships go more smoothly there than in a young immigrant country such as the United States, where there is still considerable awkwardness about greetings, appeasement gestures, and ceremony. Although various appropriate gestures are used in the United

States, they are not yet nationally institutionalized; for example: one person may offer his hand for a handshake whereas the other, not expecting this, has his hands in his pockets. It is not surprising, then, that a fairly constant anxiety surrounds interpersonal relations in this country, and newspaper columns on social etiquette are, probably as a consequence, extremely popular.

To demonstrate that greetings do indeed perform an important function, one can perform the experiment of not smiling or saying hello to those whom one normally greets. It will not be many mornings before the level of aggression has risen considerably, and yet no direct acts of anger will have been perceived. Contrariwise, the institutionalization of greeting and appeasement gestures where none before existed may lighten the social atmosphere considerably. For example, a chronically unhappy group can be made more spontaneous and warm if a person high in the hierarchy initiates greetings, smiles frequently, engages in interested chit-chat, and thereby precludes a preoccupation with feelings of anger. Social gatherings may have a similar function for groups; parties, dances, and ceremonies of various sorts can serve, among other things, to reduce the buildup of antagonisms.

Averting of eyes is usually a gesture of appeasement, and it is often necessary to keep aggression low, especially between strange males, since the direct confrontation of eyes can be taken as a challenge to one's dominance. This is an interesting phenomenon which holds over a wide range of species, and one can, for example, elicit challenge after challenge in a zoo by staring into the eyes of dominant animals; subordinate animals, by contrast, merely avert their eyes.

Averting of the eyes and head is closely related to turning the back and presenting the rear as a gesture of appeasement among primates; in a group whose members are familiar with each other such motor patterns usually serve to prevent further attack. Interestingly enough, very similar behavior is to be found on any schoolground during recess, particularly among the boys, as when one calls off the "attack" of another by turning away and crouching. The message transmitted seems to be: "I'm in no mood or not fearful enough to run from you, but I acknowledge your greater potency."

The present essay has been an attempt to sketch a non-definitive evolutionary framework for viewing man's social behavior. I have omitted discussion of many important aspects of human social behavior which seem to be present in all societies and which are thus species-specific: religion and mourning, dance, song, festivity, and humor, among others. All share the common importance of binding members of the group to one another and, as a matter of fact, functional interpretations of these behaviors are available. (For examples, see Wallace [1966] on religion; Freud [1938] on humor; Evans-Pritchard [1965] on dance.)

It seems fruitless to carry on a debate about whether these and the other behaviors we have discussed are or are not products of phylogenetic adaptation. Only the accumulation of properly designed empirical work can settle such issues, and in our laboratories we are making a start in this direction. For example, it has been proposed that the cry of a baby is phylogenetically adaptive in that it elicits caretaking reactions in human adults. As an initial study we are making a number of psycho-physiological measures of responsivity to a series of recorded sounds, including baby cries, in women who have been independently assessed as either high or low in maternal responsiveness. In this way, we hope to find physiological correlates to the caretaking response and thereby to understand it better.

In a second study, which seeks to examine experimentally the adaptive value of beardedness, we simply used spontaneous associations and judgments to bearded vs. non-bearded pictures as a first step. To eliminate the possibility of stereotyped responses, we plan as a second step to present the pictures subliminally, using a tachistoscope, so that conscious registration is avoided. In this technique the judgments and associations are made to a neutral "masking" figure instead. This method may be used with a wide variety of facial expressions purported to have phylogenetically adaptive function, such as blushing with shame, reddening with anger, the direct vs. the indirect threat-stare, and so on. In general, behavioral experimentation on human subjects performed within an evolutionary framework is practically nonexistent, despite the great possibilities, because psychologists have simply not thought in this way.

In addition to the need for empiricism, we should not lose sight of the basic logical principle which guides an evolutionary approach to man. Stated broadly, it asserts that everything which man does must, at some level, reflect his biological makeup and his evolutionary past, and that man's cultural and biological nature are two aspects of the same macrofeedback system. Thus, although it is true that societies and civilizations change with time, the conservatism of basic institutions and behavior is necessarily matched by the conservatism of man's genotype—for they are, indeed, parts of the same phenomenon, the one we call man.

Despite obvious problems with this broad approach, progress in the science of man demands an initial overall conceptualization which can serve as a reasonable guide to detailed exploration. The other alternatives (for example, that of the reflexologists) have stressed the search for basic units which, it was hoped, would by accretion end with a scientifically based conception of man; the history of biology and psychology, however, has proved the extreme limitations of such an atomistic approach (Goldstein, 1939).

Finally, as scientists, it is our hope that the experimental examination of corollaries stemming from the evolutionary point of view will soon begin to appear in scientific journals which concern themselves with the behavior of man, even as they have appeared for many years now in journals of animal behavior.

Bibliography

Chapter 1

ALLEE, W. C. 1951. *Co-operation among Animals.* New York: Henry Schuman.

ALVERDES, F. 1927. *Social Life in the Animal World.* New York: Harcourt Brace & Co.

ARMSTRONG, E. 1947. *Bird Display and Behavior.* New York: Oxford University Press.

BARTHOLOMEW, G. 1953. Behavioral factors affecting social structure in the Alaska fur seal. *In:* J. B. TREFETHEN (ed.), *Trans. Eighteenth No. Amer. Wildlife Conference.* Washington, D.C.: Wildlife Management Institute.

BURT, W. 1943. Territoriality and home range concepts as applied to mammals. *J. Mammal.,* **24:** 346–52.

CARPENTER, C. R. 1942. Characteristics of social behavior in non-human primates. *Trans. N.Y. Acad. Sci.,* Ser. II., **4:** 248–58.

———. 1958. Territoriality. *In:* A. ROE and G. G. SIMPSON (eds.), *Behavior and Evolution.* New Haven: Yale University Press.

COLLIAS, N. E. 1944. Aggressive behavior among vertebrate animals. *Physiol. Zool.,* **17:** 83–123.

———. 1951. Problems and principles of animal sociology. *In:* C. STONE (ed.), *Comparative Psychology.* New York: Prentice-Hall.

CROOK, S. 1961. The basis of flock organization in birds. *In:* W. H. THORPE and O. L. ZANGWILL (eds.), *Current Problems in Animal Behaviour.* Cambridge: Cambridge University Press.

DARLING, F. F. 1937. *A Herd of Red Deer.* London: Oxford University Press.

DOBZHANSKY, T. 1951. *Genetics and the Origin of Species.* New York: Columbia University Press.

GREENBERG, B., and NOBLE, G. K. 1944. Social behaviour of the American chameleon (*Anolis carolinensis* Voigt). *Physiol. Zool.,* **17:** 392–439.

GUHL, A. M. 1953. *Social Behavior of the Domestic Fowl* (*Tech. Bull. 73*). Kansas State College, Manhattan, Kans.: Agriculture Experiment Station.

HEDIGER, H. 1955. *Psychology of Animals in Zoos and Circuses*. New York: Criterion Books.

HOFSTADTER, R. 1944, 1955. *Social Darwinism in American Thought*. Boston: Beacon Press.

HOWARD, E. 1920. *Territory in Bird Life*. London: John Murray. Reprinted, 1948; London: William Collins Sons & Co.

KROPOTKIN, P. 1914. *Mutual Aid*. New York: Alfred A. Knopf.

LACK, D. 1943. *The Life of the Robin*. Reprinted, 1953; London: Penguin Books.

MURCHISON, C. 1935. The experimental measurement of a social hierarchy in *Gallus domesticus*, I. *J. Gen. Psychol.*, **12**: 3–39.

NICE, M. 1937. Studies in the life history of the song sparrow, I. *Trans. Linnaean Soc.* (N.Y.), **4**: 1–247.

SCHJELDERUP-EBBE, T. 1935. Social behavior in birds. *In:* C. MURCHISON (ed.), *Handbook of Social Psychology*. Worcester, Mass.: Clark University Press.

SHAW, E. 1962. The schooling of fishes. *Sci. Amer.*, June, 1962.

STEWART, J., and SCOTT, J. P. 1947. Lack of correlation between leadership and dominance relationships in a herd of goats. *J. Comp. Physiol. Psychol.*, **40**: 255–64.

TINBERGEN, N. 1953. *The Herring Gull's World*. London: William Collins Sons & Co.

Chapter 2

ALTMANN, M. 1952. Social behavior of elk, *Cervus canadensis* Nelsoni, in the Jackson Hole area of Wyoming. *Behaviour*, **4**: 116–43.

ARMSTRONG, E. 1947. *Bird Display and Behaviour*. New York: Oxford University Press.

ARONSON, L. R. 1957. Reproductive and parental behavior. *In:* M. BROWN (ed.), *The Physiology of Fishes*, Vol. 2, pp. 271–304. New York: Academic Press.

BARTHOLOMEW, G. 1952. Reproductive and social behavior of the northern elephant seal. *Univ. Calif. Publ. Zool.*, **47**: 369–472.

BIRCH, H. G., and CLARK, G. 1946. Hormonal modification of social behavior. *Psychosom. Med.*, **8**: 320–31.

BLAUVELT, H. 1955. Dynamics of the mother-newborn relationship in goats. *In:* B. SCHAFFNER (ed.), *Group Processes*. New York: Josiah Macy, Jr. Foundation.

BOURLIERE, F. 1954. *The Natural History of Mammals*. New York: Alfred A. Knopf.

BRISTOWE, W. S. 1941. *Comity of Spiders*. London: The Ray Company.

BURTON, M. 1954. *Animal Courtship*. New York: Frederick A. Praeger.

CARPENTER, C. R. 1942. Sexual behavior of free-ranging rhesus monkeys (Macaca mulatta). *J. Comp. Psychol.*, **33**: 133–62.

CLARK, E., and ARONSON, L. R. 1951. Sexual behavior in the guppy, *Lebistes reticulatus* (Peters). *Zoologica*, **36**: 49–66.

COLLIAS, N. E., and JAHN, L. R. Social behavior and breeding success in Canada geese (*Branta canadensis*) confined under semi-natural conditions. *Auk*, **76:** 418–509.

DARLING, F. 1937. *A Herd of Red Deer*. London: Oxford University Press.

———. 1952. Social behavior and survival. *Auk*, **69:** 183–91.

EMLEN, J. T. 1955. The study of behavior in birds. *In:* A. WOLFSON (ed.), *Recent Studies in Avian Biology*. Urbana, Ill.: University of Illinois Press.

GREENBERG, B., and NOBLE, G. K. 1944. Social behavior of the American chameleon (*Anolis carolinensia* Voigt). *Physiol. Zool.*, **17:** 392–439.

HEDIGER, H. 1955. *Psychology of Animals in Zoos and Circuses*. New York: Criterion Books.

HEINROTH, O., and HEINROTH, K. 1958. *The Birds*. Ann Arbor: University of Michigan Press.

HUXLEY, J. S. 1914. The courtships of the great crested grebe. *Proc. Zool. Soc. London*, **11:** 491–562.

IMANISHI, K. 1957. Social behavior in Japanese monkeys, Macaca fuscata. *Psychologia*, **1:** 47–54.

KENDEIGH, S. C. 1952. *Parental Care and Its Evolution in Birds*. Urbana, Ill.: University of Illinois Press.

LACK, D. 1953. *The Life of the Robin*. Rev. ed.; London: Penguin Books.

LEHRMAN, D. S. 1961. Hormonal regulation of parental behavior in birds and infrahuman mammals. *In:* W. C. YOUNG (ed.), *Sex and Internal Secretions*. Baltimore: Williams & Wilkins Co.

LORENZ, K. 1952. *King Solomon's Ring*. New York: Thomas Y. Crowell Co.

MARSHALL, A. 1961. Breeding seasons and migration. *In:* A. MARSHALL (ed.), *Biology and Comparative Physiology of Birds*. New York: Academic Press.

MICHENER, C., and MICHENER, M. 1951. *American Social Insects*. New York: D. Van Nostrand Co.

MURIE, A. 1944. *The Wolves of Mount McKinley*. ("Fauna of the National Parks of the U.S.," No. 5.) Washington, D.C.: Superintendent of Documents.

NICE, M. M. 1937. Studies in the life history of the song sparrow, I. *Trans. Linnaean Soc.* (N.Y.), **4:** 1–247.

NOBLE, G. K., and ARONSON, L. R. 1942. The sexual behavior of Anura, I. *Bull. Amer. Mus. Nat. Hist.*, **80:** 127–42.

NOBLE, R. 1945. *The Nature of the Beast*. New York: Doubleday Doran Co.

PYCRAFT, W. P. 1914. *The Courtship of Animals*. New York: Henry Holt & Co.

SCHNEIRLA, T. C. 1938. A theory of army-ant behavior based upon the analysis of activities in a representative species. *J. Comp. Psychol.*, **25:** 51–90.

STONOR, C. 1940. *Courtship and Display among Birds*. London: Country Life.

TAVOLGA, W. 1956. Visual, chemical and sound stimuli as cues in sex discriminatory behavior of the Gobiid fish, *Bathygobius sporator*. *Zoologica*, **41:** 49–64.

TINBERGEN, N. 1953. *The Herring Gull's World*. London: William Collins Sons & Co.

——. 1953. *Social Behaviour in Animals*. New York: John Wiley & Sons.

ZUCKERMAN, S. 1932. *Social Life of Monkeys and Apes*. New York: Harcourt, Brace & Co.

Chapter 3

ARMSTRONG, E. 1950. The nature and functions of displacement activities. *In: Physiological Mechanisms in Animal Behavior*. New York: Academic Press.

BAERENDS, G. P. 1950. Specializations in organs and movements with a releasing function. *In: Physiological Mechanisms in Animal Behavior*. New York: Academic Press.

BEACH, F. A. 1945. Current concepts of play in animals. *Amer. Nat.*, **79**: 523–41.

BLEST, A. D. 1961. The concept of ritualization. *In:* W. H. THORPE and O. L. ZANGWILL (eds.), *Current Problems in Animal Behavior*. Cambridge: Cambridge University Press.

COLLIAS, N. E. 1962. Social development in birds and mammals. *In:* E. L. BLISS (ed.), *Roots of Behavior*. New York: Harper & Bros.

CRAIG, W. 1918. Appetites and aversions as constituents of instincts. *Biol. Bull.*, **34**: 91–107.

DAANJE, A. 1950. On the locomotory movements of birds, and the intention movements derived from them. *Behavior, 3*: 48–98.

EMLEN, J. 1955. The study of behavior in birds. *In:* A. WOLFSON (ed.), *Recent Studies in Avian Biology*. Urbana, Ill.: University of Illinois Press.

GUITON, P. 1959. Socialization and imprinting in brown leghorn chicks. *Animal Behavior, 7*: 26–34.

HEDIGER, H. 1955. *Psychology of Animals in Zoos and Circuses*. New York: Criterion Books.

HESS, E. 1962. Imprinting and the "critical period" concept. *In:* E. L. BLISS (ed.), *Roots of Behavior*. New York: Harper & Bros.

HINDE, R. A. 1961. Behavior. *In:* A. J. MARSHALL (ed.), *Biology and Comparative Physiology of Birds*, Chap. 23. New York: Academic Press.

LACK, D. 1943. *The Life of the Robin*. Reprinted, 1953; London: Penguin Books.

LANYON, W. E. 1960. The ontogeny of vocalization in birds. *In:* W. E. LANYON and W. TAVOLGA (eds.), *Animal Sounds and Communication*. Washington, D.C.: American Institute of Biological Sciences.

LEHRMAN, D. S. 1953. A critique of Konrad Lorenz's theory of instinctive behavior. *Quart. Rev. Biol.*, **28**: 337–63.

LORENZ, K. Z. 1950. The comparative method in studying innate behaviour patterns. *In: Physiological Mechanisms in Animal Behavior*. New York: Academic Press.

——. 1957. Companionship in bird life and other essays. *In:* C. SCHILLER (ed.), *Instinctive Behavior*. New York: International Universities Press.

NICE, M. M. 1943. Studies in the life history of the song sparrow, II. *Trans. Linnaean Soc.* (N.Y.), **6**: 1–238.

NISSEN, H. 1951. Social behavior in primates. *In:* C. P. STONE (ed.), *Comparative Psychology*. 3d ed.; New York: Prentice-Hall.

ROWELL, C. 1961. Displacement grooming in the chaffinch. *Animal Behaviour,* **9**: 38–64.

SCHEIN, M. W., and HALE, E. B. 1959. The effect of early social experience on male sexual behaviour of androgen injected turkeys. *Animal Behaviour,* **7**: 189–200.

SCHNEIRLA, T. C. 1956. Interrelationship of the "innate" and the "acquired" in instinctive behavior. *In: L'Instinct dans de comportment des animaux et de l'homme,* pp. 383–452. Paris: Masson et Cie.

SETON, E. T. 1953. *Lives of Game Animals*. 3 vols. Newton Centre, Mass.: Charles T. Branford.

SMITH, S., and HOSKING, E. 1956. *Birds Fighting*. London: Faber & Faber.

THORPE, W. H. 1956. *Learning and Instinct in Animals*. Cambridge, Mass.: Harvard University Press.

TINBERGEN, N. 1951. *The Study of Instinct*. Oxford: Clarendon Press.

———. 1953. *Social Behaviour in Animals*. New York: John Wiley & Sons.

WHITMAN, C. O. 1919. *The Behavior of Pigeons,* ed. H. CARR. Washington, D.C.: Carnegie Institute.

YOUNG, W. C. 1961. Hormones and mating behavior. *In:* W. C. YOUNG (ed.), *Sex and Internal Secretions*. Baltimore: Williams & Wilkins Co.

Chapter 4

BIRDS

ARMSTRONG, E. 1947. *Bird Display and Behavior*. New York: Oxford University Press.

HEINROTH, O., and HEINROTH, K. 1958. *The Birds*. Ann Arbor: University of Michigan Press.

HOWARD, E. 1948. *Territory in Bird Life*. 2d ed.; London: William Collins Sons & Co.

KENDEIGH, S. C. 1952. *Parental Care and Its Evolution in Birds*. Urbana, Ill.: University of Illinois Press.

LACK, D. 1943. *The Life of the Robin*. London: H. F. & G. Witherly. Reprinted, 1953; London: Penguin Books.

NICE, M. M. 1937, 1943. Studies in the life history of the song sparrow, I, II. *Trans. Linnaean Soc.* (N.Y.), **4**: 1–246; **6**: 1–239.

TINBERGEN, N. 1939. Field observations of east Greenland birds. *Trans. Linnaean Soc.* (N.Y.), **5**: 1–91.

———. 1953. *The Herring Gull's World*. London: William Collins Sons & Co.

WHITMAN, C. O. 1919. *The Behavior of Pigeons,* ed. H. CARR. Publ. No. 257, pp. 1–161. Washington, D.C.: Carnegie Institute.

MAMMALS

BOLWIG, N. 1959. A study of the behavior of the Chacma baboon, *Papio wisinus*. *Behaviour*, 14: 136–63.

BOURLIERE, F. 1954. *The Natural History of Mammals*. New York: Alfred A. Knopf.

CALHOUN, J. B. 1948. Mortality and movement of brown rats. (*Rattus norvegicus*) in artificially supersaturated populations. *J. Wildlife Monogr.*, 13: 167–72.

———. 1952. The social aspects of population dynamics. *J. Mammal.*, 33: 139–59.

CARPENTER, C. R. 1934. A field study of the behavior and social relations of the howling monkeys. *Comp. Psychol. Monogr.*, 10: No. 2.

———. 1942. Sexual behavior of free ranging rhesus monkeys (*Macaca mulatta*). *J. Comp. Psychol.*, 33: 133–62.

DARLING, F. F. 1937. *A Herd of Red Deer*. London: Oxford University Press.

DAVIS, D., EMLEN, J., and STOKES, A. 1948. Studies on home range in the brown rat. *J. Mammal.*, 29: 207–25.

IMANISHI, K. 1957. Social behavior in Japanese monkeys, *Macaca fuscata*. *Psychologia*, 1: 47–54.

MURIE, A. 1944. *The Wolves of Mount McKinley* (Fauna of the National Parks of the U.S., No. 5). Washington, D.C.: U.S. Government Printing Office.

SCHALLER, G. 1963. *The Mountain Gorilla*. Chicago: University of Chicago Press.

ZUCKERMAN, S. 1932. *The Social Life of Monkeys and Apes*. New York: Harcourt, Brace.

HUMAN EVOLUTION

Symposia

GAVAN, J. (ed.). 1955. *The Non-Human Primates and Human Evolution*. A symposium. Detroit: Wayne University Press. (See particularly the articles by CARPENTER, NISSEN, HAYES and HAYES.)

MONTAGU, M. (ed.). 1962. *Culture and the Evolution of Man*. A symposium largely of reprinted articles. New York: Oxford University Press. (See particularly the articles by OAKLEY, WASHBURN, BARTHOLOMEW and BIRDSELL, CHANCE, ETKIN, DOBZHANSKY and MONTAGU, METTLER, HALLOWELL, EISELEY, and MONTAGU.)

ROE, A., and SIMPSON, G. G. (eds.). 1958. *Behavior and Evolution*. A symposium. Detroit: Wayne State University Press. (See particularly the articles by COLBERT, PRIBRAM, NISSEN, HARLOW, THOMPSON, WASHBURN, and AVIS, HUXLEY, FREEDMAN and ROE, and MEAD.)

SPUHLER, J. N. (ed.). 1958. *Natural Selection in Man.* A symposium. Detroit: Wayne State University Press. (See particularly the article by NEEL.)

———. 1959. *The Evolution of Man's Capacity for Culture.* A symposium. Detroit: Wayne State University Press. (See particularly the articles by SPUHLER, WASHBURN, and SAHLINS.)

TAX, SOL (ed.). 1960. *The Evolution of Man.* A symposium. Chicago: University of Chicago Press. (See particularly the articles by LEAKEY, WASHBURN and HOWELL, CRITCHLEY, and HALLOWELL.)

WASHBURN, S. (ed.). 1961. *Social Life of Early Man.* A symposium. Chicago: Aldine Publishing Co. (See particularly the articles by BOURLIERE, CHANCE, HEDIGER, SCHULTZ, WASHBURN, and DE VORE, OAKLEY, HALLOWELL and CASPARI.)

Journal Articles

BARTHOLOMEW, G., and BIRDSELL, S. 1953. Ecology and the protohominids. *American Anthropologist,* **55:** 481–98. Reprinted in A. MONTAGU (ed.), 1962. (See symposia listed above.)

CHANCE, M., and MEAD, A. 1953, 1962. Social behavior and primate evolution. *Symposia of Society for Experimental Biology,* **7:** 395–439. A revision of this article by Chance is included in A. MONTAGU (ed.), 1962. (See symposia listed above.)

DART, R. 1960. The bone tool-manufacturing ability of *Australopithecus prometheus. American Anthropologist,* **62:** 134–43.

EISELEY, L. 1956. Fossil man and human evolution. *In:* THOMAS (ed.), *Yearbook of Anthropology,* 1955. Reprinted in A. MONTAGU (ed.), 1962. (See symposia listed above.)

ETKIN, W. 1954. Social behavior and the evolution of man's mental faculties. *Amer. Nat.,* **88:** 129–42. Reprinted with additional discussion in A. MONTAGU (ed.), 1962. (See symposia listed above.)

———. 1963a. Social behavioral factors in the emergence of man. *Human Biology,* **35:** 299–311.

———. 1963b. Animal communication. *In:* J. EISENSON (ed.), *Psychology of Communication.* New York: Appleton-Century Crofts.

FISHER, J., and HINDE, R. A. 1949. The opening of milk bottles by birds. *Brit. Birds,* **42:** 347–57.

IMANISHI, K. 1961. The origin of human family (English summary). *Jap. J. Ethnology,* **25:** 119–38.

ITANI, J. 1958. On the acquisition and propagation of a new food habit in the natural group of the Japanese monkey at Takasaki Yama. *J. Primatology,* **1:** 84–99.

KAWAI, M. 1958. On the rank system in a natural group of Japanese monkeys, I. *J. Primatology,* **1:** 111–31.

KAWAI, M., and MIZUHARA, H. 1959. An ecological study on the wild mountain gorilla (*Gorilla gorilla beringei*). *J. Primatology,* **2:** 1–43.

LEAKEY, L., and DES BARTLETT. 1960. Finding the world's earliest man. *Nat. Geog.,* **118:** 420–35.

OAKLEY, K. 1957. Tools makyth man. *Antiquity*, **31**: 199.

————. 1951, 1962. A definition of man. *Science News*, 20. Reprinted in A. MONTAGU (ed.), 1962. (See symposia listed above.)

YERKES, R. M. 1943. *Chimpanzees: A Laboratory Colony*. New Haven, Conn.: Yale University Press.

WASHBURN, S. 1960, 1962. Tools and human evolution. *Sci. Amer.*, **203**: 63–75. Reprinted in A. MONTAGU (ed.), 1962. (See above and also symposia article in A. ROE and G. G. SIMPSON [eds.], 1958.)

WASHBURN, S., and DE VORE, I. 1961. Social behavior of baboons and early man. *In:* S. WASHBURN (ed.), *Social Life of Early Man*. Chicago: Aldine Pub. Co.

WEIDENREICH, F. 1947. The trend of human evolution. *Evolution*, **1**: 221–36.

Chapter 5

AINSWORTH, M. D. 1962. *The Effects of Maternal Deprivation: A Review of Findings and Controversy in the Context of Research Strategy*. Public Health Papers, 14. Geneva: World Health Organization.

AMBROSE, J. A. 1960. The smiling and related responses in early human infancy: An experimental and theoretical study of their course and significance." Ph.D. diss., University of London.

ARDREY, R. 1966. *The Territorial Imperative: A Personal Inquiry into the Animal Origins of Property and Nations*. New York: Atheneum.

ASCH, S. E. 1952. *Social Psychology*. New York: Prentice-Hall.

AUSUBEL, D. P. 1958. *Theory and Problems of Child Development*. New York: Grune and Stratton.

BAKAN, D. 1966. *The Duality of Human Existence*. Chicago: Rand McNally.

BARKOW, J. H. Causal interpretation of correlation in cross-cultural studies. *American Anthropologist* (in press).

BAYLEY, N. 1961. Personal communication.

BEACH, F. (ed.) 1965. *Sex and Behavior*. New York: Wiley.

BERNE, E. 1964. *Games People Play: The Psychology of Human Relationships*. New York: Grove Press.

BIRREN, J. E. (ed.) 1959. *Handbook of Aging and the Individual*. Chicago: University of Chicago Press.

BLURTON JONES, N. G. 1966. Some aspects of the social behaviour of children in nursery school. *In:* MORRIS, D. (ed.), *Primate Ethology*. London: Weidenfeld and Nicolson.

BOCK, R. D., and VANDENBERG, S. G. 1966. "Components of heritable variation in mental test scores." Read at Second Louisville Conference on Human Behavior Genetics.

BOOTH, P. B. 1966. "Sex differences in manifest content of South African dreams." M.A. thesis, University of Chicago.

BUBER, M. 1958. *I and Thou*. (2d ed.) New York: Scribner.

BUYTENDIJK, F. J. J. 1962. The phenomenological approach to the problem of feelings and emotions. *In:* H. RUITENBEEK (ed.), *Psychoanalysis and Existential Philosophy*. New York: E. P. Dutton & Co.

CHOMSKY, N. 1965. *Aspects of the Theory of Syntax*. Cambridge, Mass.: M.I.T. Press.

DARLINGTON, C. D. 1958. *The Evolution of Genetic Systems*. (2d ed.) Cambridge University Press.

DART, R. A. 1956. The cultural status of the South African man-apes. Smithsonian Report for 1955, pp. 317–38. Washington, D.C.: Smithsonian Institution.

DARWIN, C. 1872. *The Origin of Species*. (6th ed., New York: Collier Books, 1962.)

DARWIN, C. 1873. *The Expression of the Emotions in Man and Animals*. London: John Murray. (Reprinted University of Chicago Press, 1965).

DAVENPORT, W. 1965. Sexual patterns and their regulation in a society of the Southwest Pacific. *In:* F. A. BEACH (ed.), *Sex and Behavior*. New York: Wiley.

DEVORE, I. (ed.) 1965. *Primate Behavior: Field Studies of Monkeys and Apes*. New York: Holt, Rinehart and Winston.

ETKIN, W. 1967. Behavioral factors stabilizing social organization in animals. *In:* J. ZUBIN and H. HUNT, *Comparative Psychopathology: Animal and Human*. New York: Grune and Stratton.

EVANS-PRITCHARD, E. E. 1965. The dance. *In: The Position of Women in Primitive Societies and Other Essays in Social Anthropology*. New York: Free Press.

FANTZ, R. L. 1966. The origin of form perception. *In:* S. Coopersmith (ed.), *Frontiers of Psychological Research: Readings from Scientific American*. San Francisco: W. H. Freeman.

FELDMAN, S. 1962. Blushing, fear of blushing, and shame. *J. Amer. Psychoan. Assoc.*, **10**, 368–85.

FESTINGER, L. 1966. Cognitive dissonance. *In:* S. Coopersmith (ed.), *Frontiers of Psychological Research: Readings from Scientific American*. San Francisco: W. H. Freeman.

FINE, R. 1956. *Psychoanalytic Observations on Chess and Chess Masters*. New York: National Psychological Association for Psychoanalysis.

FORD, C. S., and BEACH, F. A. 1952. *Patterns of Sexual Behavior*. New York: Harper Bros.

FREEDMAN, D. G. 1964. Smiling in blind infants and the issue of innate vs. acquired. *J. Child Psychol. Psychiat.*, **5**, 171–84.

FREEDMAN, D. G. 1965. Hereditary control of early social behavior. *In:* B. M. Foss (ed.), *Determinants of Infant Behaviour III*. New York: Wiley.

FREEDMAN, D. G. 1967. Personality development in infancy, a biological approach. *In:* Y. BRACKBILL (ed.), *Infancy and Childhood*. New York: Free Press.

FREEDMAN, D. G., BOVERMAN, H., and FREEDMAN, N. 1966. Effects of kinesthetic stimulation on weight-gain and on smiling in premature infants."

Presented to the annual meeting of the American Orthopsychiatric Association, San Francisco. (Mimeographed.)

FREUD, S. 1922. *Group Psychology and Analysis of the Ego*. London: International Psychoanalytic Press.

FREUD, S. 1938. Wit and its relation to the unconscious. *In:* FREUD, *Basic Writings*. New York: Random House.

GIBSON, E. R., and WALK, R. D. 1960. The visual cliff. *Scientific American*, **202**, 64–71.

GLUCKMAN, M. 1963. Gossip and scandal. *Current Anthropology*, **4**, 307–16.

GOLDSTEIN, KURT. 1939. *The Organism*. New York: American Book Company.

GOODALL, J. 1965. Chimpanzees of the Gombe Stream Reserve. *In:* I. DEVORE (ed.), *Primate Behavior: Field Studies of Monkeys and Apes*. New York: Holt, Rinehart and Winston.

GOODE, W. J. 1959. The theoretical importance of love. *Amer. Soc. Rev.*, **24**, 38–47.

GREENMAN, G. W. 1963. Visual behavior of newborn infants. *In:* A. J. SOLNIT and S. A. PROVENCE, *Modern Perspectives in Child Development*. New York: International Universities Press.

GUTTMAN, R. 1965. A design for the study of the inheritance of normal mental traits. *In:* S. G. VANDERBERG (ed.), *Methods and Goals in Human Behavior Genetics*. New York: Academic Press.

HALEY, J. 1958. The art of psychoanalysis. *Etc.: A Review of General Semantics*, **15**, 190–200.

HALL, K. R. L., and DEVORE, I. 1965. Baboon social behavior. *In:* I. DEVORE (ed.), *Primate Behavior: Field Studies of Monkeys and Apes*. New York: Holt, Rinehart and Winston.

HARLOW, H. F., and HARLOW, M. 1966. Learning to love. *American Scientist*, **54**, 244–72.

HARLOW, H. F., and ZIMMERMAN, R. R. 1959. Affectional responses in the monkey. *Science*, **130**, 421–32.

HEBB, D. O. 1953. Heredity and environment in mammalian behavior. *Brit. J. of Anim. Behav.*, **1**, 43–47.

HUXLEY, J. S. 1958. Cultural process and evolution. *In:* A. ROE and G. G. SIMPSON (eds.), *Behavior and Evolution*. New Haven: Yale University Press.

JERSILD, A. T. 1954. Emotional development. *In:* L. CARMICHAEL (ed.), *Manual of Child Psyhcology*. New York: Wiley.

JOLLY, A. 1967. *Lemur behavior*. Chicago: University of Chicago Press.

KINSEY, A. C. 1953. *Sexual Behavior in the Human Female*. Philadelphia: W. B. Saunders.

KINSEY, A. C., POMEROY, W. B., and MARTIN, C. E. 1948. *Sexual Behavior in the Human Male*. Philadelphia: W. B. Saunders.

KOFORD, C. B. 1963. Group relations in an island colony of rhesus monkeys. *In:* C. H. SOUTHWICK (ed.), *Primate Social Behavior*. New York: Van Nostrand.

KOHLBERG, L. 1963. The development of children's orientations toward a moral order: I. Sequences in the development of moral thought. *Vita Humana*, **6**, 11–33.

KORTLANDT, A. and KOOIJ, M. 1963. Protohominid behaviour in primates. *In:* Symposia of the Zoological Society of London, No. 10.

KOVACH, J. K., and HESS, E. H. 1963. Imprinting: Effects of painful stimulation upon the following response. *J. Comp. Physiol. Psychol.*, **56**, 461–64.

LEVINE, R. 1965. Socialization, social structure, and intersocietal images. *In:* H. C. KELMAN (ed.), *International Behavior: A Social-Psychological Analysis.* New York: Holt, Rinehart and Winston.

LEVINE, R. 1967. Personal communication.

LINTON, R. 1939. Marquesan culture. *In:* A. KARDINER, *The Individual and His Society: The Psychodynamics of Primitive Social Organization.* New York: Columbia University Press.

LORENZ, K. 1958. The evolution of behavior. *Scientific American*, **199**: 67–83.

LORENZ, K. 1965. *Evolution and Modification of Behavior.* Chicago: University of Chicago Press.

LORENZ, K. 1966. *On Aggression.* New York: Harcourt, Brace and World.

MACCOBY, E. (ed.) 1966. *The Development of Sex Differences.* Stanford, Calif.: Stanford Press.

MALINOWSKI, B. 1956. What is a family? *In:* M. A. MONTAGU (ed.), *Marriage: Past and Present.* Boston: Porter Sargent.

MASTERS, W. H., and JOHNSON, V. E .1965. The sexual response cycles of the human male and female: Comparative anatomy and physiology. *In:* F. A. BEACH (ed.), *Sex and Behavior.* New York: Wiley.

MEAD, M. 1939. *From the South Seas: Studies of Adolescence and Sex in Primitive Societies.* New York: W. Morrow and Co.

MONTAGU, M. A. 1946. *Adolescent Sterility: A Study in the Comparative Physiology of the Infecundity of the Adolescent Organism in Mammals and Man.* Springfield, Ill. C. C Thomas.

MURDOCK, G. P. 1957. World ethnographic sample. *Amer. Anthrop.*, **59**, 664–87.

NEUGARTEN, B. 1966. Adult personality: Toward a psychology of the life-cycle. Presented to American Psychological Association. New York.

OSGOOD, C. E., MIRON, M. S., and ARCHER, W. K. 1963. The cross-cultural generality of effective meaning systems: Progress report. Center for Comparative Psycholinguistics, University of Illinois.

PIAGET, J. 1950. *The Moral Judgment of the Child.* Glencoe, Ill.: Free Press.

ROHEIM, G. 1950. *Psychoanalysis and Anthropology: Culture, Personality and the Unconscious.* New York: International Universities Press.

SAINT-EXUPÉRY, A. DE. 1943. *The Little Prince.* New York: Harcourt, Brace.

SAWYER, J., and GUETZKOW, H. 1965. Bargaining and negotiation in international relations. *In:* H. C. KELMAN (ed.), *International Behavior: A Social-Psychological Analysis.* New York: Holt, Rinehart and Winston.

SCHUMPETER, J. A. 1951. *Imperialism and Social Classes.* New York: A. M. KELLEY.

SIMPSON, G. G. 1944. *Tempo and Mode in Evolution.* New York: Columbia University Press.

SPIETH, H. T. 1952. Mating behavior within the genus Drosophila (Diptera). *Bull. of Am. Mus. of Natural Hist.,* Vol. 99, Art. 7, New York.

SUGGS, R. C. 1966. *Marquesan Sexual Behavior.* New York: Harcourt, Brace and World.

SUMNER, W. G. 1906. *Folkways: A Study of the Sociological Importance of Usages, Manners, Customs, Mores, and Morals.* (Reprinted New American Library, 1960).

TANNER, J. M. 1961. *Education and Physical Growth.* London: University of London Press.

TINBERGEN, N. 1953. *The Herring Gull's World: A Study of the Social Behaviour of Birds.* London: Collins.

UEXKÜLL, J. VON. 1926. *Theoretical Biology.* New York: Harcourt, Brace.

VAYDA, A. P. 1961. Expansion and warfare among swidden agriculturalists. *American Anthropologist,* **63,** 346–58.

WADDINGTON, C. H. 1960. *The Ethical Animal.* New York: Atheneum.

WALLACE, A. F. C. 1966. *Religion: An Anthropological View.* New York: Random House.

WHITE, L. A. 1949. Energy and the evolution of culture. *In: The Science of Culture,* 363–93.

WILSON, EDMUND. 1955. *To the Finland Station: A Study in the Writing and Acting of History.* Garden City, N. Y. Doubleday.

WOLFF, P. H. 1963. Observations on the early development of smiling. *In:* B. M. Foss (ed.), *Determinants of Infant Behavior,* II. London: Methuen.

WYNNE-EDWARDS, V. C. 1962. *Animal Dispersion in Relation to Social Behavior.* New York: Hafner.

YOUNG, W. C. 1965. The organization of sexual behavior by hormonal action during the prenatal and larval periods in vertebrates. *In:* F. BEACH (ed.), *Sex and Behavior.* New York: Wiley.

ZUCKERMAN, S. 1932. *The Social Life of Monkeys and Apes.* London: Routledge and Kegan Paul.

Index

PHOENIX BOOKS
in Science

PHOENIX SCIEN

...h· An Introduction to

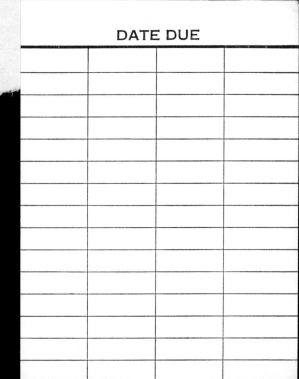

DATE DUE
